A Short Introduction to Preferences

Between Artificial Intelligence and Social Choice

Synthesis Lectures on Artificial Intelligence and Machine Learning

Editors

Ronald J. Brachman, *Yahoo Research*
William W. Cohen, *Carnegie Mellon University*
Thomas Dietterich, *Oregon State University*

A Short Introduction to Preferences: Between Artificial Intelligence and Social Choice
Francesca Rossi, Kristen Brent Venable, and Toby Walsh
2011

Human Computation
Edith Law and Luis von Ahn
2011

Trading Agents
Michael P. Wellman
2011

Visual Object Recognition
Kristen Grauman and Bastian Leibe
2011

Learning with Support Vector Machines
Colin Campbell and Yiming Ying
2011

Algorithms for Reinforcement Learning
Csaba Szepesvári
2010

Data Integration: The Relational Logic Approach
Michael Genesereth
2010

Markov Logic: An Interface Layer for Artificial Intelligence
Pedro Domingos and Daniel Lowd
2009

Introduction to Semi-Supervised Learning
XiaojinZhu and Andrew B.Goldberg
2009

Action Programming Languages
Michael Thielscher
2008

Representation Discovery using Harmonic Analysis
Sridhar Mahadevan
2008

Essentials of Game Theory: A Concise Multidisciplinary Introduction
Kevin Leyton-Brown and Yoav Shoham
2008

A Concise Introduction to Multiagent Systems and Distributed Artificial Intelligence
Nikos Vlassis
2007

Intelligent Autonomous Robotics: A Robot Soccer Case Study
Peter Stone
2007

A Short Introduction to Preferences: Between Artificial Intelligence and Social Choice
Francesca Rossi, Kristen Brent Venable, and Toby Walsh

ISBN: 978-3-031-00428-5   paperback
ISBN: 978-3-031-01556-4   ebook

DOI 10.1007/978-3-031-01556-4

A Publication in the Springer series
SYNTHESIS LECTURES ON ARTIFICIAL INTELLIGENCE AND MACHINE LEARNING

Lecture #14
Series Editors: Ronald J. Brachman, *Yahoo Research*
                William W. Cohen, *Carnegie Mellon University*
                Thomas Dietterich, *Oregon State University*
Series ISSN
Synthesis Lectures on Artificial Intelligence and Machine Learning
Print 1939-4608   Electronic 1939-4616

A Short Introduction to Preferences

Between Artificial Intelligence and Social Choice

Francesca Rossi
University of Padova

Kristen Brent Venable
University of Padova

Toby Walsh
NICTA and University of New South Wales

SYNTHESIS LECTURES ON ARTIFICIAL INTELLIGENCE AND MACHINE LEARNING #14

ABSTRACT

Computational social choice is an expanding field that merges classical topics like economics and voting theory with more modern topics like artificial intelligence, multiagent systems, and computational complexity. This book provides a concise introduction to the main research lines in this field, covering aspects such as preference modelling, uncertainty reasoning, social choice, stable matching, and computational aspects of preference aggregation and manipulation.

The book is centered around the notion of preference reasoning, both in the single-agent and the multi-agent setting. It presents the main approaches to modeling and reasoning with preferences, with particular attention to two popular and powerful formalisms, soft constraints and CP-nets. The authors consider preference elicitation and various forms of uncertainty in soft constraints. They review the most relevant results in voting, with special attention to computational social choice. Finally, the book considers preferences in matching problems.

The book is intended for students and researchers who may be interested in an introduction to preference reasoning and multi-agent preference aggregation, and who want to know the basic notions and results in computational social choice.

KEYWORDS

preference and constraint reasoning, multiagent systems, computational social choice, collective decision making, stable matching

Contents

Acknowledgments

We are very grateful to Craig Boutilier, Ulrich Endriss and the anonymous reviewers for their constructive comments on draft versions of this book. We would also like to thank our editor Mike Morgan for encouraging us to write this book and publishing it.

Francesca Rossi and K. Brent Venable are partially supported by the MIUR PRIN 20089M932N project on "Innovative and multi-disciplinary approaches for constraint and preference reasoning". Toby Walsh's employer, NICTA, is funded by the Australian Government's Department of Broadband, Communications and the Digital Economy and the Australian Research Council through the ICT Centre of Excellence program. Toby Walsh also receives support from the Asian Office of Aerospace Research and Development (AOARD-104123).

Francesca Rossi, K. Brent Venable, and Toby Walsh
July 2011

CHAPTER 1

Introduction

Preferences are a common feature of everyday decision making. They are, therefore, an essential ingredient in many reasoning tools. Preferences are often used in collective decision making when multiple agents need to choose one out of a set of possible decisions: each agent expresses its preferences over the possible decisions, and a centralized system aggregates such preferences to determine the "winning" decision. Consequently, preference reasoning and multi-agent preference aggregation are areas of growing interest within Artificial Intelligence (AI).

Preferences are also the subject of study in social choice, especially in the area of elections and voting theory. In an election, the voters express their preferences over the candidates and a voting rule is used to elect the winning candidate. Economists, political theorist, mathematicians, as well as philosophers have invested considerable effort in studying this scenario and have obtained many theoretical results about the desirable properties of the voting rules that one can use.

Since the voting setting is closely related to multi-agent decision making, it is not surprising that in recent years the area of multi-agent systems has witnessed a growing interest in trying to reuse social choice results in the multi-agent setting. However, it soon became clear that an adaptation of such results is necessary, since several issues, which are typical of multi-agent settings and AI scenarios, usually do not occur, or have a smaller impact, in typical voting situations.

In a multi-agent system, the set of candidates can be very large with respect to the set of voters. Usually in social choice, it is the opposite: there are many voters and a small number of candidates. Also, in many AI scenarios, the candidates often have a combinatorial structure. That is, they are defined via a combination of features. Moreover, the preferences over the features are often dependent on each other. In social choice, usually the candidates are tokens with no structure. In addition, for multi-issue elections, the issues are usually independent of each other.

This combinatorial structure allows for the compact modelling of the preferences over the candidates. Therefore, several formalisms have been developed in AI to model such preference orderings. In social choice, little emphasis is put on how to model preferences, since there are few candidates, so one can usually explicitly specify a linear order.

In AI, a preference ordering is not necessarily linear, but it may include indifference and incomparability. Moreover, often uncertainty is present, for example in the form of missing or imprecise preferences. In social choice, usually all preferences are assumed to be present, and a preference order over all the candidates is a linear order that is explicitly given as a list of candidates.

Finally, multi-agent systems must consider the computational properties of the system. In social choice this usually has not been not a crucial issue.

It is therefore very interesting to study how these two disciplines, social choice and AI, can fruitfully cooperate to give innovative and improved solutions to aggregating preferences of multiple agents. We will try to give an account of the interdisciplinary nature of research in this area, while providing a glimpse of all the issues mentioned above.

In Chapter 2, we start by presenting the main approaches to modeling and reasoning with preferences, with particular attention to soft constraints and CP-nets. Then, in Chapter 3, we consider preference elicitation and various forms of uncertainty given by missing, imprecise, or vague preferences. In Chapter 4, we consider multi-agent preference aggregation and review the most relevant results in social choice as well as their computational aspects. Finally, in Chapter 5, we consider matching problems, a large class of problems where agents express preferences over each other.

Most of the material described in this book was presented at a tutorial on preferences at IJCAI 2009. The tutorial, as well as this book, is meant for students and researchers who are interested in an introduction to preference modelling and reasoning. The book is also targeted at researchers from within other areas of AI interested in understanding the applications of preferences to their area. No specific prerequisite knowledge is essential. However, some general knowledge of AI and computational complexity is advisable.

CHAPTER 2

Preference Modeling and Reasoning

Representing and reasoning about preferences is an area of increasing theoretical and practical interest in AI. Preferences and constraints occur in real-life problems in many forms. Intuitively, constraints are restrictions on the possible scenarios: for a scenario to be feasible, all constraints must be satisfied. For example, if we want to buy a PC, we may impose a lower limit on the size of its screen: only PCs that respect this limit will be considered. Preferences, on the other hand, express desires, satisfaction levels, rejection degrees, or costs. For example, we may prefer a tablet PC to a regular laptop, we may desire having a webcam, as well as spending as little as possible. In this case, all PCs will be considered, but some will be preferred to others.

In many real-life optimization problems, we may have both constraints and preferences. For example, in a product configuration problem [171], the producer may impose some constraints (e.g., component compatibility) as well as preferences in the forms of optimization criteria (e.g., minimize the supply time), or also subjective preferences over alternative products expressed in some language of preference statements (e.g., pairwise comparisons).

Preferences and constraints are closely related notions, since preferences can be seen as a form of "relaxed" constraints. For this reason, there are several constraint-based preference modeling frameworks. One of the most general of such frameworks defines a notion of *soft constraints* [134], which extends the classical constraint formalism to model preferences in a *quantitative* way, by expressing several degrees of satisfaction that can be either totally or partially ordered. The term *soft constraints* is used to distinguish this kind of constraints from the classical ones, that are usually called *hard constraint*. However, hard constraints can be seen as an instance of the concept of soft constraints where there are just two levels of satisfaction. In fact, a hard constraint can only be satisfied or violated, while a soft constraint can be satisfied at several levels. When there are both levels of satisfaction and levels of rejection, preferences are usually called *bipolar*, and they can be modeled by extending the soft constraint formalism [21].

Preferences can also be modeled in a *qualitative* (also called *ordinal*) way, that is, by pairwise comparisons. In this case, soft constraints (or their extensions) are not suitable. However, other AI preference formalisms are able to express preferences qualitatively, such as CP-nets [24]. More precisely, CP-nets provide an intuitive way to specify conditional preference statements that state the preferences over the instances of a certain feature, possibly depending on some other features. For example, we may say that we prefer a red car to a blue car if the car is a sports car. CP-nets and soft constraints can be combined, providing a single environment where both qualitative and quantitative preferences can be modeled and handled.

Specific types of preferences come with their own reasoning methods. For example, *temporal preferences* are quantitative preferences that pertain to the position and duration of events in time. Soft constraints can be embedded naturally in a temporal constraint framework to handle this kind of preference [116, 144].

While soft constraints generalize the classical constraint formalism providing a way to model several kinds of preferences, this added expressive power comes at a cost, both in the modeling task as well as in the solution process. To mitigate these drawbacks, various AI techniques have been adopted. For example, *abstraction theory* [45] has been exploited to simplify the process of finding a most preferred solution of a soft constraint problem [18]. By abstracting a given preference problem, the idea is to obtain a new problem with a kind of preferences which is simpler to handle by the available solver, such that the solution of this simpler problem may give some useful information for the solution process of the original problem. Also, *inference* and *explanation computation* has been applied to preference-based systems to ease the understanding of the result of the solving process. In the context of soft constraints, explaining the result of a solving process means justifying why a certain solution is the best that can be obtained, or why no solution has been found, possibly suggesting minimal changes to the preferences that would provide the given problem with some solutions. For example, explanations have been used to guide users of preference-based configurators [75].

On the modeling side, it may be too tedious or demanding of a user to specify all the soft constraints. *Machine learning techniques* have therefore been used to learn the missing preferences [161, 185]. Alternatively, *preference elicitation* techniques [34], interleaved with search and propagation, have been exploited to minimize the user's effort in specifying the problem while still being able to find a most preferred solution [85].

2.1   CONSTRAINT REASONING

Constraint programming [48, 163] is a powerful paradigm for solving combinatorial search problems that draws on a wide range of techniques from artificial intelligence, computer science, databases, programming languages, and operations research. Constraint programming is currently applied with success to many domains, such as scheduling, planning, vehicle routing, configuration, networks, and bioinformatics. The basic idea in constraint programming is that the user states the constraints and a general purpose *constraint solver* is used to solve them.

2.1.1   CONSTRAINTS

Constraints are just relations, and a *constraint satisfaction problem* (CSP) states which relations should hold among the given decision variables. For example, in scheduling activities in a company, the decision variables might be the starting times and the durations of the activities and the resources needed to perform them, and the constraints might be on the availability of the resources and on their use by a limited number of activities at a time. Another example is configuration, where constraints are used to model compatibility requirements among components or user's requirements. For example, if we were to configure a laptop, some video boards may be incompatible with certain monitors. Also, the user may impose constraints on the weight and/or the screen size.

Formally, a constraint satisfaction problem (CSP) can be defined by a triple $\langle X, D, C \rangle$, where X is a set of variables, D is a collection of domains, as many as the variables, and C is a set of constraints. Each constraint involves some of the variables in X and a subset of the Cartesian product of their domains. This subset specifies the combination of values of the variables (of the constraint) that satisfy the constraint. A solution for a constraint satisfaction problem is an assignment of all the variables to values in their respective domains such that all constraints are satisfied.

As an example, consider the CSP with four variables $\{x_1, \ldots, x_4\}$, domain $\{1, 2, 3\}$ for all variables, and constraints $x_1 \neq x_2$, $x_2 \neq x_3$, $x_3 \neq x_4$, and $x_2 \neq x_4$. This is usually called a graph coloring problem, since the values in the variable domains can be interpreted as colors and the constraints say that some pairs of variables should have a different color. This CSP has several solutions. One of them is $\langle x_1 = 1, x_2 = 2, x_3 = 3, x_4 = 1 \rangle$: if we assign the variables in this way, all constraints are satisfied. CSPs can be graphically represented by a *constraint graph* where nodes model variables and arcs (or hyperarcs) model constraints. The constraint graph of this CSP is shown in Figure 2.1.

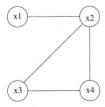

Figure 2.1:   The constraint graph of a constraint satisfaction problem.

2.1.2   CONSTRAINT SOLVERS

Constraint solvers take a problem, represented in terms of decision variables and constraints, and find one or more assignments to all the variables that satisfy the constraints. In general, finding such an assignment is NP-hard and is computationally difficult. However, there are *islands of tractability*: for example, if the constraint graph does not have cycles, then finding a solution requires just polynomial time in the size of the CSP [49, 74].

Constraint solvers search the solution space either systematically, as with *backtracking* or *branch-and-bound* algorithms, or they use forms of *local search*, which may be incomplete. The state space of the problem can be represented by a tree, called the *search tree*. Each node in such a tree represents an assignment to some of the variables. The root represents the empty variable assignment, while each leaf represents an assignment to all the variables. Given a node, its children are generated by selecting a variable not yet assigned and choosing as many assignments for this variable as the number of values in its variable domain. Each arc connecting a node with its children is labelled with one of those values, meaning that this value is assigned to the selected variable. Each path in the search tree (from the root to a node) represents a (possibly incomplete) variable assignment. Since each node has a unique path from the root to it, there is a one-to-one correspondence between tree nodes and variable assignments.

Consider again the graph coloring example above. Its search tree has the root node, plus 3 nodes at the first level (for the 3 possible ways to instantiate the first variables), 9 nodes at the second level, 27 nodes at the third level, and 81 leaves.

Backtracking search.   Backtracking search performs a partial depth-first traversal of the search tree. At each step, a new variable is instantiated with a value in its domain, unless there are no values which are compatible with the previous variable assignments. Compatibility here means that the new variable assignment, together with the previous ones, satisfies all the constraints among such variables. We stop when all variables are instantiated, or report failure when all possible instantiations for the variables have been considered and some variables remain uninstantiated. The time complexity of backtracking search is exponential in the number of variables, while its space complexity is linear.

For the graph coloring problem above, if we choose to assign variables in increasing order of their index, and to try domain values from the smallest to the largest, backtracking search would visit the part of the search tree shown in Figure 2.2 a) before finding the first solution of the CSP. The arrows denote the sequence of steps made by the algorithm to visit some nodes of the search tree, assuming it always follows the leftmost alternative. In this case, no backtracking is necessary to find a solution. If instead we have just two colors (thus two values in each variable domain, say 1 and 2), then backtracking search will visit the part of the search tree shown in Figure 2.2 b) before reporting a failure, meaning that the problem has no solution.

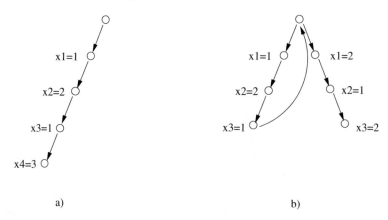

a)   b)

Figure 2.2:   Backtracking search for the CSP of Fig. 2.1. Part a) considers three colors per variable, while part b) considers just two colors.

Local search.   Local search [104] is an alternative approach for solving computationally hard combinatorial problems. Given a problem instance, the basic idea underlying local search is to start from an initial search position in the space of all possible assignments (typically, a randomly or heuristically generated assignment, which may be infeasible, sub-optimal or incomplete), and to improve iteratively this assignment by means of minor modifications. At each search step, we move to a new assignment selected from a local neighborhood via an heuristic evaluation function. For example, the neighborhood could contain assignments that differ in the value assigned to one variable, and the evaluation function could be the number of satisfied constraints. This process is repeated until a solution is found or a predetermined number of steps is reached. To ensure that the search process does not stagnate, most local search methods make use of random moves: for example, at every step, with a certain probability a random move might be performed rather than the usual move to the best neighbor.

Branch-and-bound search.   Rather than trying to satisfy a set of constraints, finding any variable assignment that does not violate any of them, we may want to distinguish among such feasible assignments, and select one that is optimal according to some optimization criterion. Such criteria are usually modeled via an objective function that measures the quality of each solution. The aim is then to find a solution with optimal (say, maximal) quality, where the quality of a solution can be expressed in terms of preferences. For such problems, branch-and-bound is often used to find an optimal solution.

Branch-and-bound search traverses the search tree like backtracking search, except that it does not stop when one solution is found, but it continues looking for possibly better solutions. To do this, it keeps two values during the tree visit: the quality of the best solution found so far (initialized with a dummy solution with the worst possible value) and, given the current node t, an over-estimation of the quality of all complete assignments below t (usually called an *heuristic function*). When such an over-estimation is worse than the quality of the best solution found so far, we can avoid visiting the subtree rooted at t since it cannot contain any solution which is better than the best one found so far. After exhaustively visiting the nodes in this search tree, the last solution stored is optimal according to the objective function.

Consider again the CSP in Fig. 2.1, with three colors for each variable. Suppose the objective function is to maximize $x_1 + x_3$. Let us assume we use the heuristic function $max(D_1) + max(D_3)$, where D_1 and D_3 are the domains of x_1 and x_3, respectively. Then, assuming we try variables and values in the same order as above, branch-and-bound search would visit the part of the search tree shown in Fig. 2.3 before finding an optimal solution, with a value of the objective function of 6. The arrows denote the sequence of steps made by the algorithm to visit the nodes of the search tree, assuming it always follows the leftmost alternative. For example, the algorithm first instantiates all variables down the leftmost branch (thus obtaining the solution $\langle x_1 = 1, x_2 = 2, x_3 = 1, x_4 = 3\rangle$), then it jumps back to a new instantiation for x_3, and finds a second solution $\langle x_1 = 1, x_2 = 2, x_3 = 3, x_4 = 1\rangle$), then it jumps back to the root, and so on. Each node, except the leaves, has a certain value of the heuristic functions h, while the leaves, which represent complete variable assignments, are associated with a certain value of the objective function f.

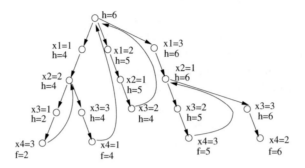

Figure 2.3:   Branch-and-bound search for the CSP of Fig. 2.1.

Constraint propagation.   Constraint propagation [16] involves modifying the constraints or the variable domains, for example by deleting values from the domains, without changing the solution set. This can be seen as a form of inference. For example, constraint propagation can be used to infer that some values can be eliminated from the domain of a particular variable, based on the other domains and on the constraints, since these values cannot be part of any solution.

The most widely used form of constraint propagation is *arc consistency*. A constraint between two variables x and y is arc-consistent if, for every value a in the domain of x there exists a value b in the domain of y such that the assignment $(x = a, y = b)$ satisfies the constraint, and vice versa.

Given any constraint satisfaction problem, there are many polynomial-time algorithms to achieve arc consistency over all the constraints in the problem. At the end, we have an new problem with the same variables and constraints, but possibly smaller variable domains. Moreover, the new problem has the same set of solutions as the original one.

Consider a constraint between variables x and y, which states that x should be smaller than y, and assume the domain of both variables is $\{1, 2, 3\}$. This constraint is not arc consistent, since when $x = 3$, there is no value in the domain of y which, together with $x = 3$, would satisfy the constraint. Also, if $y = 1$, there is no value for x which, together with $y = 1$, satisfies the constraint. We can make the constraint arc consistent by eliminating 3 from the domain of x and 1 from the domain of y.

Since a backtracking or a branch-and-bound algorithm searching for a solution need to consider the different possible instantiations for the variables from their domains, the search space is smaller if the domains are smaller. Thus, achieving arc consistency (or, more generally, performing some form of constraint propagation) may greatly help the behaviour of the search algorithms by reducing the parts of the search space that need to be visited. Systematic solving methods therefore often interleave search and constraint propagation.

2.2   SOFT CONSTRAINTS

While constraint satisfaction methods have been successfully applied to many real-life combinatorial problems, in some cases the hard constraint framework is not expressive enough. For example, it is possible that after having listed the desired constraints on the decision variables, there is no way to satisfy them all. In this case, the problem is said to be *over-constrained*, and the model needs to be refined to relax some of the constraints. This relaxing process, even when it is feasible, is rarely formalized and is normally difficult and time consuming. Even when all the constraints can be satisfied, we may want to discriminate between the (equally good) solutions. These scenarios often occur when constraints are used to formalize desired properties rather than requirements that cannot be violated. Such desired properties are not faithfully represented by constraints, but they should rather be considered as *preferences* whose violation should be avoided as much as possible.

As an example, consider a typical university timetabling problem which aims at assigning courses and teachers to classrooms and time slots. There are usually many constraints, such as the size of the classrooms, the opening hours of the building, or the fact that the same teacher cannot teach two different classes at the same time. However, there are usually also many preferences, which state, for example, the desires of the teachers (like that he prefers not to teach on Fridays), or university policies (like that it is preferable to use smaller classrooms). If all these preferences are modeled by constraints, it is easy to find scenarios where there is no way to satisfy all of them. In this case, what one would like is to satisfy all hard requirements while violating the preferences as little as possible. Thus, preferences must be represented as different from constraints. Modeling preferences correctly also allows us to discriminate among all the solutions which satisfy the hard constraints. In fact, there could be two timetables which both satisfy the hard requirements, but where one of them better satisfies the desires, and this should be the chosen one. Similar scenarios can be found in most of the typical application areas for constraints, such as scheduling, resource allocation, rostering, vehicle routing, etc.

In general, preferences can be *quantitative* or *qualitative* (e.g., "I like this at level 10" versus "I like this more than that"). Preferences can also be conditional (e.g., "If the main dish is fish, I prefer white wine to red wine"). Preferences and constraints may also co-exist. For example, in a product configuration problem, there may be production constraints (for example, a limited number of convertible cars can be built each month), marketing preferences (for example, that it would be better to sell the standard paint types), whilst the user may have preferences of various kind (for example, that if it is a sport car, she prefers red).

To cope with some of these scenarios, hard constraints have been generalized in various ways in the past decades. The underlying observation of such generalizations is that hard constraints are relations, and thus they can either be satisfied or violated. Preferences need instead a notion that has several levels of satisfiability. In the early '90s, several *soft constraint* formalisms were proposed that generalize the notion of constraint to allow for more than two levels of satisfiability.

2.2.1   SPECIFIC SOFT CONSTRAINT FORMALISMS

Here we list some specific soft constraint formalisms, among which fuzzy constraints, weighted constraints, and probabilistic constraints.

Fuzzy constraints.   Fuzzy constraints (see, for instance, [59, 170]) use (discretized) preference values between 0 and 1. The quality of a solution is the minimum preference associated with constraints in that solution. The aim is then to find a solution whose quality is highest. Since only the minimum preference is considered when evaluating a variable assignment, fuzzy constraints suffer from the so-called "drowning effect" (that is, the worst level of satisfiability "drowns" all the others). This is typical of a pessimistic approach, that can be useful or even necessary in critical applications, such as in the medical or aerospace domain. However, this approach is too pessimistic in other domains. For this reason, *lexicographic constraints* were introduced [70] to obtain a more discriminating ordering of the solutions: to order two solutions, we compare lexicographically the ordered sequence of all the preference values given by the constraints to those two solutions. In this way, solutions with different minimum preference values are ordered as in the classical fuzzy constraint setting, but also solutions with the same minimum preference (that would be equally preferred in fuzzy constraints) can be discriminated.

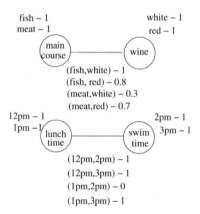

Figure 2.4:   A fuzzy constraint problem.

An example of a fuzzy constraint problem can be seen in Figure 2.4. We are deciding on what to have for lunch and when to go swimming. The fuzzy CSP has four variables (represented by the circles), each with two values. For example, wine can be either red or white. There are three constraints, represented by solid arrows, and giving a preference value between 0 and 1 to each assignment of the variables of the constraint. If we consider the variable assignment ⟨main course = meat, wine = white, lunch time = 1pm, swim time = 2pm⟩, then its preference value is 0, since we have to take the minimum contribution of all constraints, and the constraint between lunch time and swim time gives preference value 0. If instead we consider the variable assignment ⟨main course

= fish, wine = white, lunch time = 12pm, swim time = 2pm), then its preference value is 1, since all constraints are satisfied at level 1. Thus, this second assignment is preferred to the first one. Since the scale of preference values is between 0 and 1, with 1 modeling the highest preference, this second assignment is also optimal.

Probabilistic constraints.   Another extension of hard constraints are *probabilistic constraints* [69], where, in the context of an uncertain model of the real world, each constraint is associated with the probability of being present in the real problem. Solutions are then associated with their conjoint probability (assuming independence of the constraints), and the aim is to find a solution with the highest probability.

Weighted constraints.   In many real-life problems, it is often natural to talk about costs or penalties, rather than preferences. In fact, there are situations where we are more interested in the damage caused by not satisfying a constraint rather than in the advantage we obtain when we satisfy it. For example, when we want to use tolled highways to get to a different city, we may want to find the itinerary with the minimum overall cost. Also, when we want to put together a PC by buying all of its parts, we want to find the combination of parts which has the minimum price. A natural way to extend the hard constraint formalism to deal with these situations consists of associating a certain penalty or cost to each constraint, to be paid when the constraint is violated.

In *weighted constraints*, each constraint is associated with a weight, and the aim is to find a solution for which the sum of the weights of the satisfied constraints is maximal. A very useful instance of weighted constraints are MaxCSPs, where weights are just 0 or 1 (0 if the constraint is violated and 1 if it is satisfied). In this case, we want to satisfy as many constraints as possible.

It is easy to transform probabilities into additive costs by taking their logarithm, and this allows us to reduce any probabilistic constraint instance to a weighted constraint instance [174]. Notice, however, that probabilistic constraints are similar to fuzzy constraints, since in both cases the values associated to the constraints are between 0 and 1, and better solutions have higher values. The main difference is that, while in fuzzy constraints the evaluation of a solution is the minimum value (over all the constraints), in probabilistic constraints it is the product of all the values.

Weighted constraints are among the most expressive soft constraint frameworks, in the sense that the task of finding an optimal solution for possibilistic, lexicographic or probabilistic frameworks can be efficiently (that is, in polynomial time) reduced to the task of finding an optimal solution for a weighted constraint instance [174].

While fuzzy, lexicographic, and probabilistic constraints were defined to model real-life situations that could not be faithfully modeled via hard constraints, weighted constraints and MaxCSPs have mainly been defined to address over-constrained problems, where there are so many constraints that the problem has no solution. In this case, the goal is to satisfy as many constraints as possible, possibly giving them some importance levels.

2.2.2 GENERAL SOFT CONSTRAINT FORMALISMS

Partial constraint satisfaction.   Over-constrained problems have been the motivation also for the definitions of the first general framework to extend hard constraints in order to model preferences, called *partial constraint satisfaction* [78]. In order to find a solution for an over-constrained hard constraint problem, partial constraint satisfaction tries to identify another hard constraint problem which is both consistent and as *close* as possible to the original. The space of problems considered to find this consistent network is defined using constraint relaxations (by, for example, forgetting constraints or adding extra authorized combinations to the existing constraints) together with a specific metric, which is needed to identify the nearest problem. MaxCSPs are then just an instance of partial CSPs where the metric is simply the number of satisfied constraints.

Semiring-based soft constraints.   Another general constraint-based formalism for modeling preferences is the *semiring-based formalism* [19, 134], which encompasses most of the previous extensions. Its aim is to provide a single environment where properties of specific preference formalisms (such as fuzzy or weighted constraints) can be proven once and for all, and then inherited by all the instances. At the technical level, this is done by introducing a structure representing the levels of satisfiability of the constraints. Such a structure is just a set with two operations: one (written +) is used to generate an ordering over the levels, while the other one (×) is used to define how two levels can be combined and which level is the result of such combination. Moreover, two elements of the set are the *best* and *worst* element among all preference levels. Because of the properties required on such operations, this structure can be defined as a variant of the notion of semiring. More precisely: a *c-semiring* is a 5-tuple $\langle E, +_s, \times_s, \mathbf{0}, \mathbf{1} \rangle$ such that:

- E is a set containing at least $\mathbf{0}$ and $\mathbf{1}$;

- $+_s$ is a binary operator closed in E, associative, commutative and idempotent, for which $\mathbf{0}$ is a neutral element and $\mathbf{1}$ an annihilator;

- $\times_s$ is a binary operator closed in E, associative and commutative, for which $\mathbf{0}$ is an annihilator and $\mathbf{1}$ a neutral element;

- $\times_s$ distributes over $+_s$.

In words, the minimum level $\mathbf{0}$ is used to capture the notion of absolute non-satisfaction, which is typical of hard constraints. Since a single complete unsatisfaction is unacceptable, $\mathbf{0}$ must be an annihilator for $\times_s$. This means that, when combining a completely violated constraint with a constraint which is satisfied at some level, we get a complete violation. Conversely, a complete satisfaction should not hamper the overall satisfaction degree, which explains why $\mathbf{1}$ is a neutral element for $\times_s$. In fact, this means that, when we combine a completely satisfied constraint and a constraint which is satisfied at some level l, we get exactly l. Moreover, since the overall satisfaction should not depend on the way elementary satisfactions are combined, combination (that is, $\times_s$) is required to be commutative and associative.

To define the ordering over the preference levels, operator $+_s$ is used: if $a +_s b = b$, it means that b is preferred to a, and we write this as $b \succeq_s a$. If $a +_s b = c$, and c is different from both a and b, a and b are incomparable. To make sure that this ordering has the right properties, operator $+_s$ is required to be associative, commutative and idempotent. This generates a partial order, and more precisely a lattice. In all cases, $a +_s b$ is the least upper bound of a and b in the lattice $< E, \succeq_s >$. The fact that $\mathbf{1}$ (resp. $\mathbf{0}$) is a neutral (respectively, annihilator) element for $+_s$ follows from the fact that it is a maximum (respectively, minimum) element for $\succeq_s$.

Finally, assume that a is better than b, and consider two complete assignments, one that satisfies a constraint at level a and the other one that satisfies the same constraint at level b. Then, if all the other constraints are satisfied equally by the two assignments, it is reasonable to expect that the assignment satisfying at level a is overall better than the one satisfying at level b. For comparable a and b, this is equivalent to saying that $\times_s$ distributes over $+_s$. In a c-semiring, this property is required in all cases, even if a and b are incomparable.

This gives the *semiring-based soft constraint problems* (SCSPs), where constraints have several levels of satisfiability, that are (totally or partially) ordered according to the semiring structure. More formally, a *semiring constraint problem* is a tuple $\langle X, D, C, S \rangle$ where:

- $X = \{x_1, \ldots, x_n\}$ is a finite set of n variables.

- $D = \{D_1, \ldots, D_n\}$ is the collection of the domains of the variables in X such that D_i is the domain of x_i.

- $S = \langle E, +_s, \times_s, \mathbf{0}, \mathbf{1} \rangle$ is a c-semiring.

- C is a finite set of soft constraints. A soft constraint is a function f on a set of variables $V \subseteq X$, called the scope of the constraint, such that f maps assignments (of variables in V to values in their domains) to semiring values, that is, $f : \prod_{x_i \in V} D_i \to E$. Thus, a soft constraint can be viewed as a pair $\langle f, V \rangle$ also written as f_V.

Fuzzy, lexicographic, probabilistic, weighted, and MaxCSPs are all instances of the semiring-based framework. Indeed, even hard constraints are an instance of the semiring-based framework. With hard constraints, are either satisfied or violated, and this can be modeled by having two levels of preferences (true and false). Moreover, a complete variable assignment is a solution of a CSP when all constraints are satisfied, and this can be modeled by choosing the logical and operator to combine constraints. The logical or instead models the ordering between the two preference levels, ensuring that true is better than false. Thus, for hard constraints, the c-semiring is $= \langle \{false, true\}, \vee, \wedge, false, true \rangle$. For fuzzy constraints, the c-semiring is $\langle [0, 1], max, min, 0, 1 \rangle$. In fact, preference levels are between 0 and 1, with 1 being the best and 0 being the worst value (which is achieved by choosing the max operator to define the ordering over preference levels), and preference combination is performed via the min operator (only the worst preference value is used to assess the quality of a variable assignment).

Valued constraints.   Similar to semiring-based constraints, *valued constraints* were also introduced as a general formalism to model constraints with several levels of satisfiability [174]. Valued constraints are indeed very similar to semiring-based soft constraints, the main difference being that the preference values cannot be partially ordered [20].

The possibility of partially ordered preference values can be useful is several scenarios. For example, when the preference values are tuples of elements (such as when we need to combine several optimization criteria), it is natural to have a Pareto-like approach in combining such criteria, and this leads to a partial order. Also, even if we have just one optimization criterion, we may want to insist on declaring some preference values as incomparable because of what they model. In fact, the elements of the semiring structure do not need to be numbers, but can be any kind of object, that we associate to variable assignments. If, for example, they are all the subsets of a certain set, then we can have a partial order over such items under subset inclusion.

2.2.3   COMPUTATIONAL PROPERTIES OF SOFT CONSTRAINTS

Solving a semiring-based problem is a difficult task since semiring-based constraints properly generalize hard constraints, where the problem is already NP-hard. However, as with hard constraints, there are easy cases also for soft constraints such as when the constraint graph does not have cycles [47].

If the computation of $a \times_s b$ and $a +_s b$ need polynomial time in the size of their arguments (that is, a and b), then deciding if the consistency level of a problem is higher than a given threshold is an NP-complete task. Under the same conditions, given two solutions of a soft constraint problem, checking whether one is preferable to the other is polynomial: we simply compute the desirability values of the two solutions and compare them in the preference order.

Many search techniques have been developed to solve specific classes of soft constraints, like fuzzy or weighted constraints. However, all have an exponential worst-case complexity. Systematic approaches like backtracking search and constraint propagation can be adapted to soft constraints. For example, branch-and-bound can use the preference values of a subset of the constraints to compute the bound for pruning the search tree. Also, constraint propagation techniques like arc consistency can be generalized to certain classes of soft constraints, including fuzzy and weighted constraints [134].

2.2.4   BIPOLAR PREFERENCES

Bipolarity is an important topic in several fields, such as psychology and multi-criteria decision making, and it has recently attracted interest in the AI community, especially in argumentation [5], qualitative reasoning [57, 58], and decision theory [120]. Bipolarity in preference reasoning can be seen as allowing one to state both degrees of satisfaction (that is, *positive* preferences) and degrees of rejection (that is, *negative* preferences).

Positive and negative preferences can be thought as symmetric concepts, and thus one might try to deal with them using the same operators. However, this may not model what one usually expects in real scenarios. For example, if we have a dinner menu with fish and white wine, and we like them both, then having both should be more preferred than having just one of them. On the other hand, if we don't like any of them, then the preference of having them both should be smaller than the preferences of having each of them alone. In other words, the combination of positive preferences should produce a higher (positive) preference, while the combination of negative preferences should give us a lower (negative) preference. Thus, we need operators with different properties to combine positive and negative preferences.

When dealing with both kinds of preferences, it is natural to express also *indifference*, which means that we express neither a positive nor a negative preference. For example, we may say that we like peaches, we don't like bananas, and we are indifferent to apples. Then, a desired behavior of indifference is that, when combined with any preference (either positive or negative), it should not influence the overall preference. For example, if we like peaches and we are indifferent to apples, a dish with peaches and apples should have overall a positive preference.

Moreover, we also want to be able to combine positive with negative preferences. The most natural and intuitive way to do so is to allow for *compensation*. Comparing positive against negative aspects and compensating them with respect to their strength is one of the core features of decision-making processes, and it is, undoubtedly, a tactic universally applied to solve many real life problems.

Positive and negative preferences might seem as just two different criteria to reason with, and thus techniques such as those usually adopted by multi-criteria optimization [61], such as Pareto-like approaches, could appear suitable for dealing with them. However, this interpretation would hide the fundamental nature of bipolar preferences, that is, positive preferences are naturally the opposite of negative preferences.

Semiring-based constraints can model only negative preferences, since in this framework preference combination returns lower preferences. However, they have been generalized to model both positive and negative preferences [21], as well as indifference and preference compensation. This is done by adding to the usual c-semiring structure another algebraic structure to model positive preferences, and by setting the highest negative preference to coincide with the lowest positive preference, to link the two structures and to model indifference. To find optimal solutions of bipolar preference problems, it is possible to adapt the notions of soft constraint propagation and branch-and-bound search.

Bipolarity has also been considered in qualitative preference reasoning [14, 15], where fuzzy preferences model the positive knowledge and negative preferences are interpreted as violations of constraints. Precedence is given to negative preference optimization, and positive preferences are used to distinguish among the optimal solutions found in the first phase, thus not allowing for compensation. Another approach [95] considers totally ordered unipolar and bipolar preference scales and defines an operator, the *uninorm*, which can be seen as a restricted form of compensation.

2.3 CP-NETS

Soft constraints are one of the main tools for representing and reasoning about preferences in constraint satisfaction problems. However, they require specifying a semiring value for each variable assignment in each constraint. In many applications, it may be more natural for users to express preferences via generic qualitative (usually partial) preference relations over variable assignments. For example, it is often more intuitive for the user to state "I prefer red wine to white wine", rather than "Red wine has preference 0.7 and white wine has preference 0.4" (with the assumption that a higher preference value expresses higher desirability). Although the former statement provides us with less information, it does not require the careful selection of preference values for (possibly partial) variable assignments, which is instead required in soft constraints.

CP-nets [23, 24] (that is, Conditional Preference networks) are graphical models for representing and reasoning about certain types of qualitative preference statements, interpreted under the *ceteris paribus (cp)* assumption. For instance, under the ceteris paribus interpretation, the statement *"I prefer red wine to white wine if meat is served"* asserts that, given two meals that differ *only* in the kind of wine served *and* both containing meat, the meal with a red wine is preferred to the meal with a white wine. Observe that this interpretation corresponds to a "least committing" interpretation of the information provided by the user, and many philosophers (see [100] for an overview) and AI researchers [55] have argued for this interpretation of preference assertions. To emphasize the ceteris paribus interpretation, such statements are usually called cp-statements.

Informally, each CP-net compactly captures the preference relation induced by a set of such (possibly conditional) cp-statements. Structurally, CP-nets bear some similarity to Bayesian networks [118], as both utilize directed graphs where each node stands for a variable (usually called a *feature* in the CP-net literature). Thus, there is a set of features $\{X_1, \ldots, X_n\}$ with finite, discrete domains $\mathcal{D}(X_1), \ldots, \mathcal{D}(X_n)$, which play the same role as the variables in soft constraints. Another similarity between CP-nets and Bayesian networks is that graphical structure in both models relies on a notion of independence between the variables: Bayesian nets utilize the notion of probabilistic independence, while CP-nets utilize the notion of preferential independence [71].

2.3.1 CONDITIONAL PREFERENCES

During preference elicitation, for each feature X_i the user is asked to specify a set of *parent* features $Pa(X_i)$, the values of which affect her preferences over the values of X_i. This information is used to create the directed *dependency graph* of the CP-net in which each node X_i has $Pa(X_i)$ as its immediate predecessors. Given this structural information, the user is asked to specify explicitly her preference over the values of X_i for *each complete assignment* of $Pa(X_i)$, and this preference is assumed to take the form of a total [23] or partial [24] order over $\mathcal{D}(X_i)$. These conditional preferences over the values of X_i are captured by a *conditional preference table $CPT(X_i)$*, which is annotated with the node X_i in the CP-net. Each cp-preference statement has therefore the form $x_1 = v_1, \ldots, x_n = v_n : y = w_1 \succ \ldots \succ y = w_k$, where y is the variable for which we are specifying our preferences, $\{w_1, \ldots, w_k\}$ is the domain of y, $x_1, \ldots, x_n$ are the parents of y (that is, the variables on which y depends), and each v_i is in the domain of x_i, for $i = 1, \ldots, n$. Here $\succ$ means *preferred to*.

As an example, consider the CP-net in Figure 2.5 a). This CP-net has three features, which are graphically represented by nodes, of which two (*main course* and *fruit*) do not depend on any other feature, while *wine* depends on *main course*. The dependency is graphically represented by the arrow. For the independent features, we just have a total order over their values (such as *fish* $\succ$ *meat* for *main course*), while for the dependent feature we have a total order for each assignment of the feature it depends upon (thus one for *main course* = *fish* and another one for *main course* = *meat*). These total orders are shown close to the node representing the feature. Thus, this CP-net includes the following cp-statements: *fish is preferred to meat; peaches are preferred to strawberries; white wine is preferred to red wine if fish is served; otherwise, red wine is preferred to white wine.*

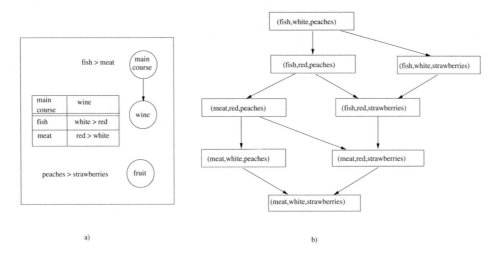

a)   b)

Figure 2.5: A CP-net and its outcome ordering.

2.3.2 PREFERENCE ORDERING

The ceteris paribus interpretation. A CP-net induces an ordering over the variable assignments, which is based on the *ceteris paribus* interpretation of the conditional preference statements. *Ceteris paribus* in Latin means *all else being equal*. This is intended to mean that two variable assignments that differ for the value of one variable, while all else is the same, are always ordered (that is, one is preferred to the other one). To understand which is preferred to which, we just have to look at the CP-table of the changing variable. For example, in the CP-net of Figure 2.5 a), assignment $\langle fish, white, peaches \rangle$ is preferred to $\langle fish, red, peaches \rangle$ since the CP-table of variable *wine* tells us that, when there is fish, we prefer white wine to red wine. In order to compute the entire ordering over all variable assignments, we use the concept of a *worsening flip*.

Worsening flips.   A worsening flip is a change in the value of a single feature to a value which is less preferred according to a cp-statement for that feature. For example, in the CP-net in Figure 2.5 a), moving from ⟨*fish,red,peaches* ⟩ to ⟨*meat,red,peaches* ⟩ is a worsening flip, since, according to the CP table for the main course, fish is preferred to meat. Another worsening flip is moving from ⟨*fish,white,berries* ⟩ to ⟨*fish,red,berries* ⟩, since, according to the CP table of *wine*, white is preferred to red in the presence of fish.

Outcome ordering.   In CP-nets, complete assignments of all the features to values in their domains are called *outcomes*. An outcome α is preferred to an outcome β, written $\alpha \succ \beta$, if and only if there is a chain of worsening flips from α to β. This definition induces a strict preorder over the outcomes, which defines the so-called *induced graph*, where nodes represent outcomes and directed arcs represent worsening flips to [24]. An outcome is optimal if there is no other outcome which is preferred to it in this preorder. Notice that CP-nets cannot model any preorder over the outcomes: for example, two outcomes that differ in the value of only one feature are always ordered (that is, they are never incomparable).

Figure 2.5 b) shows the outcome preference ordering for the CP-net in Figure 2.5 a). Outcomes appearing higher in the figure are more preferred. The preference ordering is the transitive closure of the relation given by the arrows of the figure. For example, we can see that the outcome ⟨*fish, red, strawberries* ⟩ is preferred to ⟨*meat, white, strawberries*⟩ since there is a chain of worsening flips from the first outcome to the second. On the other hand, outcomes ⟨*fish, red, peaches* ⟩ and ⟨*fish, white, berries* ⟩ are incomparable since there is no chain of worsening flips from any one to the other. We can also see that outcome ⟨*fish, white, peaches* ⟩ is the optimal outcome.

2.3.3   COMPUTATIONAL PROPERTIES

From the point of view of reasoning about preferences, several types of queries can be answered using CP-nets. First, given a CP-net N one might be interested in finding an optimal assignment to the features of N. In general, finding the optimal outcome of a CP-net is NP-hard. However, in acyclic CP-nets, there is only one optimal outcome and this can be found in linear time [23]: we simply sweep through the CP-net, following the arrows in the dependency graph and assigning at each step the most preferred value in the current feature's preference table. Second, one might be interested in comparing two assignments. Determining if one outcome is preferred to another (that is, a dominance query) is PSPACE-complete for acyclic CP-nets, even if feature domains have just two values [94]: intuitively, to check whether an outcome is preferred to another one, one should find a chain of worsening flips, and these chains can be exponentially long.

Various extensions of the notion of CP-net have been defined and studied over the years. For example, TCP-nets introduce the notion of trade-offs in CP-nets based on the idea that variables may have different importance levels [27]. Also, efficient dominance testing have been defined for preference languages more general than CP-nets [191].

2.4   SOFT CONSTRAINTS VS. CP-NETS

It is clear, at this point, that soft constraints and CP-nets have complementary advantages and drawbacks. CP-nets allow us to represent conditional and qualitative preferences, but dominance testing is expensive. On the other hand, soft constraints allow us to represent both hard constraints and quantitative preferences, and to have a cheap dominance test. We will now formally compare their expressive power and investigate whether one formalism can be approximated by the other one.

2.4.1   EXPRESSIVENESS COMPARISON

We say that a formalism B is at least as expressive as a formalism A if and only if from a problem P expressed using A it is possible to build in polynomial time (in the size of P) a problem expressed using B such that the optimal solutions are the same.

If we apply this definition to classes of soft constraints, we get, for example, that fuzzy CSPs and weighted CSPs are at least as expressive as hard constraints. If instead we use it to compare CP-nets and various instances of soft constraints, we see that hard constraints are at least as expressive as CP-nets. In fact, given any CP-net, we can obtain in polynomial time a set of constraints whose solutions are the optimal outcomes of the CP-net [26]: it is enough to take, for each preference statement of the form $x_1 = v_1, \ldots, x_n = v_n : y = w_1 \succ \ldots \succ y = w_k$, the constraint $x_1 = v_1, \ldots, x_n = v_n \Rightarrow y = w_1$. On the contrary, there are some hard constraint problems for which it is not possible to build in polynomial time a CP-net with the same set of optimals. This means that, if we are just interested in the set of optimal solutions, hard constraints are at least as expressive as CP-nets.

However, we can be more fine-grained in the comparison, and say that a formalism B is at least as expressive as a formalism A if and only if from a problem P expressed using A it is possible to build in polynomial time (in the size of P) a problem expressed using B such that the orderings over solutions are the same. Here not only we must maintain the set of optimal solutions, but also the rest of the ordering over the solutions. In this case, CP-nets and soft constraints are incomparable, since each can do something that the other one cannot do. More precisely, dominance testing (that is, comparing two complete assignments to decide which is preferred (if either)) is a difficult task in CP-nets [52]. On the contrary, it is polynomial in soft constraints (if we assume $\times_s$ and $+_s$ to be polynomially computable). Thus, unless P=NP, it is not possible to generate in polynomial time a soft constraint network which has the same solution ordering of a given CP-net. On the other hand, given any soft constraint network, it is not always possible to generate a CP-net with the same ordering. This follows from the fact that CP-nets cannot represent all possible preorders, but only some of them. For example, as we noted above, no CP-net can generate an outcome ordering where two solutions differing in just one flip are unordered. On the other hand, soft constraint networks can represent any partial order over solutions. Thus, when we are interested in the solution ordering, CP-nets and soft constraints are incomparable. This continues to hold also when we augment the CP-nets with set of hard constraints for filtering out infeasible assignments.

2.4.2   APPROXIMATING CP-NETS VIA SOFT CONSTRAINTS

It is possible to approximate a CP-net ordering via soft constraints, achieving tractability while sacrificing precision to some degree. Different approximations can be characterized by how much of the original ordering they preserve, the time complexity of generating the approximation, and the time complexity of comparing outcomes in the approximation. It is often desirable that such approximations are information preserving; that is, what is ordered in the given ordering is also ordered in the same way in the approximation. In this way, when two outcomes are ordered in the approximation, we can infer that they are either ordered or incomparable in the CP-net. If instead two outcomes are in a tie or incomparable in the approximation, they are surely incomparable in the CP-net. Another desirable property of approximations is that they preserve the ceteris paribus property. CP-nets can be approximated by both weighted and fuzzy constraints. In both cases, the approximation is information preserving and satisfies the ceteris paribus property [53].

2.4.3   CP-NETS AND HARD CONSTRAINTS

CP-nets provide a very intuitive framework to specify conditional qualitative preferences, but they do not allow for hard constraints. However, real-life problems often contain both constraints and preferences. While hard constraints and quantitative preferences can be adequately modeled and handled within the soft constraint formalisms, when we have both hard constraints and qualitative preferences, we need to either extend the CP-net formalism [22] or at least develop specific solution algorithms. Notice that reasoning with both constraints and preferences is difficult as often the most preferred outcome is not feasible, and not all feasible outcomes are equally preferred. When we have preferences expressed via a CP-net and a set of hard constraints, it is however possible to obtain all the optimal outcomes by solving a set of hard *optimality constraints*. While, in general, the proposed algorithm needs to perform expensive dominance tests, there are special cases in which this can be avoided [157].

2.5   TEMPORAL PREFERENCES

Preference formalisms can be adapted to deal with specific kinds of preferences. For example, temporal preferences lets us express preferences over the time and duration of events. Soft constraints have been used to model such temporal preferences. Reasoning about time is a core issue in many real-life problems, such as planning and scheduling for production plants, transportation, and space missions. Several approaches have been proposed to reason about temporal information. The formalisms and approaches based on *temporal constraints* have been among the most successful in practice.

In temporal constraint problems, variables either represent instantaneous events, such has "when a plane takes off", or temporal intervals, such as "the duration of the flight". Temporal constraints allow us to put temporal restrictions either on when a given event should occur, e.g., "the plane must take off before 10am", or on how long a given activity should last, e.g., "the flight should not last more than two hours".

Several quantitative and qualitative constraint-based temporal formalisms have been proposed, stemming from pioneering works by Allen [4] and by Dechter, Meiri, and Pearl [50]. A qualitative temporal constraint defines which temporal relations, (e.g., *before, after, during*), are allowed between two temporal events or intervals, representing two activities. For example, one could say "fueling must be completed before boarding the plane". The quantitative version of this constraint would instead be "the time difference between the end of the fueling task and the start of the boarding must be between 5 and 20 minutes". Disjunctions can be expressed in both formalisms. "I will call you either before or after the flight", or "I'll be flying home either between 2pm and 4pm or between 6pm and 8pm".

Once the temporal constraints have been stated, the goal is to find an occurrence time, or duration, for all the events respecting all the constraints. In general, solving temporal constraint problems is difficult. However, there are tractable classes, such as quantitative temporal constraint problems where there is only one temporal interval for each constraint, called *simple temporal constraints* [50].

The expressive power of classical temporal constraints may be insufficient to model faithfully all the aspects of the problem. For example, while it may be true that "the plane must take off before 10am", one may want to say that "the earlier the plane takes off, the better". Moreover, one may want to state that "the flight should take no more than two hours, but the ideal time is actually one and a half hours". It is easy to see that these statements are common in the specification of most temporal problems arising in practice. Both qualitative and quantitative temporal reasoning formalisms have been extended with *quantitative preferences* to allow for the specification of such types of statements.

More precisely, Allen's approach has been augmented with fuzzy preferences [9] that allow one to express soft constraints such as "I will call you either before or after the flight, and my preference for calling you before the flight is 0.2, while that for calling you after the flight is 0.9". In other words, fuzzy preferences are associated with the relations among temporal intervals allowed by the constraints. Higher values represent the fact that such a relation is more preferred. Once such constraints have been stated, then, as usual with fuzzy preferences, the goal is to find a temporal assignment to all the variables with the highest overall preference, where the overall preference of an assignment is the lowest preference given by the temporal constraints on any constraint. Such problems are solved by exploiting specific properties of fuzzy preferences, in order to decompose the optimization problem, reducing it to solving a set of hard constraint problems.

Fuzzy preferences have been combined with both disjunctive and simple quantitative temporal constraints [115, 144]. The result is the notion of soft temporal constraints, where each allowed duration or occurrence time for a temporal event is associated with a fuzzy preference representing the desirability of that specific time. For example, it is possible to state "the time difference between the end of fueling and the start of boarding must be between 5 and 20 minutes, and the preference for 5 to 10 minutes is 1 while the preferences from 11 to 20 minutes decrease linearly from 0.8 to 0.1". The decomposition approach is the most efficient solution technique also in the quantitative setting [115, 116, 144].

The main differences between the two formalisms are the objects with which preferences are associated. In the qualitative model, preferences are associated with qualitative relations between intervals, while in the quantitative model they are associated with the explicit durations or occurrence times.

Quantitative temporal constraints have also been extended with *utilitarian* preferences: preferences take values in the set of positive reals where the goal is to maximize their sum. Such problems have been solved using branch-and-bound techniques, as well as SAT and weighted constraint satisfaction approaches [138, 145].

2.6   ABSTRACTING, EXPLAINING, AND ELICITING PREFERENCES

In constraint satisfaction problems, we look for a solution, while in soft constraint problems, we look for an optimal solution. Not surprisingly, soft constraint problems are typically more difficult to solve. To ease this difficulty, several AI techniques have been used. Here we discuss just two of them: abstraction and explanation generation. Abstraction works on a simplified version of the given problem, thus hoping to have a significantly smaller search space, while explanation generation helps understand the result of the solver: it is not always easy for a user to understand why no better solution is returned.

An added difficulty in dealing with soft constraints is related to the *modeling phase*, where a user has to understand how to faithfully model his real-life problem via soft constraints. In many cases, we may end up with a soft constraint problem where some preferences are missing, for example because they are too costly to be computed or because of privacy reasons. To reason in such scenarios, we may use techniques like machine learning and preference elicitation to solve the given problem.

2.6.1   ABSTRACTION TECHNIQUES

As noted above, soft constraints are more expressive than hard constraints, but it is also more difficult to model and solve a soft constraint problem. Therefore, sometimes it may be too costly to find all, or even a single, optimal solution. Also, although traditional propagation techniques like arc consistency can be extended to soft constraints [134], such techniques can be too costly, depending on the size and structure of the partial order associated to the problem. Finally, sometimes we may not have a solver for the class of soft constraints we need to solve, while we may have a solver for another "simpler" class of soft constraints.

For these reasons, it may be reasonable to work on a simplified version of the given soft constraint problem, while trying not to lose too much information. Such a simplified version can be defined by means of the notion of abstraction, which takes a soft CSP and returns a new one which is simpler to solve. Here, as in many other works on abstraction, "simpler" may mean many things, like the fact that a certain solution algorithm finds a solution, or an optimal solution, in fewer steps, or also that the abstracted problem can be processed by machinery which is not available for the given problem (such as when we abstract from fuzzy to hard constraints and we only have a solver for hard constraints).

To define an abstraction, we can use the theory of *Galois insertions* [45], which provides a formal approach to model the simplification of a mathematical structure with its operators. Given a soft CSP (the *concrete* one), we may get an abstract soft CSP by simplifying the associated semiring, and relating the two structures (the concrete and the abstract one) via a Galois insertion. This way of abstracting constraint problems does not change the structure of the problem (the set of variables remains the same, as well as the set of constraints), but only the semiring values to be associated with the tuples of values for the variables in each constraint [18].

Abstraction has also been used also to simplify the solution process of hard constraint problems [126]. Also, the notion of value interchangeability has been exploited to support abstraction and reformulation of hard constraint problems [77]. In the case of hard constraints, abstracting a constraint problem means dealing with fewer variables and smaller domains.

Once we obtain some information about the abstracted version of a problem, we can bring back to the original problem some (or possibly all) of the information derived in the abstract context, and then continue the solution process on the transformed problem, which is a equivalent to the original. The hope is that, by following this route, we get to the final goal faster than just solving the original problem.

2.6.2   EXPLANATION GENERATION

One of the most important features of problem solving in an interactive setting is the capacity of the system to provide the user with justifications, or explanations, for its operations. Such justifications are especially useful when the user is interested in what happens at any time during search because he/she can alter features of the problem to ease the problem solving process.

Basically, the aim of an explanation is to show clearly why a system acted in a certain way after certain events. Explanations have been used for hard constraint problems, especially in the context of over-constrained problems [6, 112, 113], to understand why the problem does not have a solution and what can be modified in order to get one. In soft constraint problems, explanations also take preferences into account, and they provide a way to understand, for example, why there is no way to get a better solution.

In addition to providing explanations, interactive systems should be able to show the consequences, or implications, of an action to the user, which may be useful in deciding which choice to make next. In this way, they can provide a sort of "what-if" kind of reasoning, which guides the user towards good future choices. Fortunately, in soft constraint problems, this capacity can be implemented with the same machinery that is used to give explanations.

A typical example of an interactive system where constraints and preferences may be present, and where explanations can be very useful, are configurators [171]. A configurator is a system which interacts with a user to help him/her to configure a product. A product can be seen as a set of component types, where each type corresponds to a certain finite number of specific components, and a set of compatibility constraints among subsets of the component types. A user configures a product by choosing a specific component for each component type, such that all the compatibility constraints as well as the preferences are satisfied. For example, in a car configuration problem, a user may prefer red cars but may also not want to completely rule out other colors.

Constraint-based technology is currently used in many configurators to both model and solve configuration problems [129]: component types are represented by variables, with a value for each concrete component, and both compatibility and personal constraints are represented as constraints (or soft constraints) over subsets of such variables. User choices during the interaction with the configurator are usually restricted to specifying unary constraints, in which a certain value is selected for a variable.

Whenever a choice is made, the corresponding (unary) constraint can be added to existing compatibility and personal constraints. Constraint propagation techniques like, for example, arc consistency [163] can then rule out (some of the) future choices that are not compatible with the current choice. While providing justifications based on search is difficult, arc consistency has been used as a source of guidance for justifications, and it has been exploited to help the users in some of the scenarios mentioned above. For example, it has been shown that arc consistency enforcement can be used to provide both justifications for choice elimination and also for conflict resolution [76].

This sort of approach has also been used for preference-based configurators, using the soft version of arc consistency, whose application may decrease the preferences in some constraints (while maintaining the same semantics overall). Explanations can then describe why the preferences for some variable values decrease, and they suggest at the same time which assignments can be retracted in order to get a more preferred solution [75].

Configurators with soft constraints help users not only avoid conflicts or make the next choice so that fewer later choices are eliminated, but also get to an optimal (or good enough) solution. More precisely, when the user is about to make a new choice for a component type, the configurator shows the consequences of such a choice in terms of conflicts generated, elimination of subsequent choices, and also quality of the solutions. In this way, the user can make a choice which leads to no conflict, and which presents a good compromise between choice elimination and solution quality.

2.6.3 LEARNING AND PREFERENCE ELICITATION

In a soft constraint problem, sometimes we may know a user's preferences over some of the solutions, but have no idea on how to encode this knowledge into the constraints of the problem. That is, we have a global idea about the quality of a solution, but we do not know the contribution of individual constraints to such a measure. In such a situation, it is difficult both to associate a preference with other solutions in a compatible way and to understand the importance of each tuple and/or constraint. In other situations, we may have just a rough estimate of the preferences, either for the tuples or for the constraints. Such a scenario has been addressed [161] by using machine learning techniques based on gradient descent. More precisely, it is assumed that the level of preference for some solutions (that is, the *examples*) is known, and a suitable learning technique is defined to learn, from these examples, values to be associated with each constraint tuple, in a way that is compatible with the examples. Soft constraint learning has also been embedded in a general interactive constraint framework, where users can state both usual preferences over constraints and also preferences over solutions proposed by the system [162]. In this way, the modeling and the solution process are heavily interleaved. Moreover, the two tasks can be attacked incrementally, by specifying a few preferences at a time, and obtaining better and better solutions at each step. In this way, one requires fewer examples, since they are not given by the user all at the beginning of the solution process, but they are guided by the solver, that proposes the next best solutions and asks the user to give a feedback on them. Other approaches to learning soft constraints that provide a unifying framework for soft CSP learning have recently been developed [185]. Moreover, soft constraint learning has been exploited in the application domain of processing and interpreting satellite images [135].

Machine learning techniques have also been used to learn hard constraints (that is, to learn allowed and forbidden variable instantiations). For example, an interactive technique based on a hypothesis space containing the possible constraints to learn has been used to help the user formulate his constraints [141].

Preference learning is currently a very active area within machine learning and data mining (see [80] for a recent collection of survey and technical papers on this subject). The aim is to learn information about the preferences of an individual or a class of individuals, starting from some observations which reveal part of this information. Preference elicitation is thus a crucial ingredient in this task [34], since one would like to elicit as much information as possible without spending too many system or users' resources. An important special case of preference learning is learning how to rank: the goal here is to predict preferences in the form of total orders of a set of alternatives. The learnt preferences are often used for preference prediction, that is, to predict the preferences of a new individual or of the same individual in a new situation. This connects preference learning to several application domains, such as collaborative filtering [103, 177] and recommender systems [2, 133, 173].

2.7   OTHER PREFERENCE MODELING FRAMEWORKS

There are several other ways to model preferences, besides soft constraints and CP-nets. We will briefly describe some of them in this section.

Max SAT.   SAT is the problem of deciding if a set of clauses in propositional logic can be satisfied. Each clause is a disjunction of literals, and each literal is either a variable or a negated variable. For example, a clause can be $x \vee not(y) \vee not(z)$. Satisfying a clause means giving values (either *true* or *false*) to its variables such that the clause has value *true*. MaxSAT is the problem of maximizing the number of satisfied clauses.

Since the satisfiability problem in propositional logic (SAT) is a subcase of the constraint satisfaction problem using Boolean variables and clauses, the problem MAXSAT is clearly a particular case of the weighted constraint satisfaction problem [46, 102].

Weighted and prioritized goals. An intuitive way to express preferences consists of providing a set of goals, each of which is a propositional formula, possibly adding also extra information such as priorities or weights (see [122] for a survey of goal-based approaches to preference representation). Candidates in this setting are variable assignments, which may satisfy or violate each goal. A *weighted goal* is a propositional logic formula plus a real-valued weight. The utility of a candidates is then computed by collecting the weights of satisfied and violated goals, and then aggregating them. Often only violated goals count, and their utilities are aggregated with functions such as sum or maximin. In other cases, we may sum the weights of the satisfied goals, or we may take their maximum weight. Any restriction we may impose on the goals or the weights, and any choice of an aggregation function, give a different language. Such languages may have drastically different properties in terms of their expressivity, succinctness, and computational complexity [182].

Sometimes weights are replaced by a *priority relation* among the goals. Often with prioritized goals candidates are evaluated via the so-called *discrimin* ordering: a candidate x is better than another candidate y when, for each goal g satisfied by y and violated by x, there is a goal g' satisfied by x and violated by y such that g' has priority over g.

Bayesian networks. We already mentioned Bayesian networks because of their similarities with CP-nets. Bayesian networks [143] can also be considered as specific soft constraint problems where the constraints are conditional probability tables (satisfying extra properties) using [0, 1] as the semiring values, multiplication as $\times_s$ and the usual total ordering on [0, 1]. The Most Probable Explanation (MPE) task is then equivalent to looking for an optimal solution on such problems.

Bidding languages. Combinatorial auctions are auctions where there is a set of goods for sale. Potential buyers can make bids for subsets of this set of goods, and the auctioneer chooses which bids to accept. Bidding languages are languages used by the bidders to express their bids, that is, their preferences, to the auctioneer. In this context, usually a preference structure is given by a (monotonic) function mapping sets of goods to prices. In the OR/XOR family of bidding languages [140], bids are expressed as combinations of atomic bids of the form $\langle S, p \rangle$, where p is the price the bidder is willing to pay for the set of goods S. In the OR language, the valuation of a set is the maximal value that can be obtained by taking the sum over disjoint bids for subsets of the set. For example, the bid

$$\langle \{a\}, 2 \rangle \, OR \langle \{b\}, 2 \rangle \, OR \langle \{c\}, 3 \rangle \, OR \langle \{a, b\}, 5 \rangle$$

means that the bidders is willing to pay 2 for a or b alone, 3 for c alone, 5 for both a and b, and 8 for the whole set. In the XOR language, atomic bids are mutually exclusive, so the valuation of a set is the highest value offered for any of its subsets. The XOR language is more expressive than the OR language but is not very succinct since it may require the enumeration of all subsets with non-zero valuation. It is possible to combine these two languages to obtain bidding languages with better expressiveness and succinctness properties than either of the two. Many other concise bidding languages for combinatorial auctions have been defined and studied, which improve on OR/XOR languages (see, for example, the *generalized logical bids* (GLBs) in [25] and the *tree-based bidding language* (TBBL) in [32]).

Utility-based models.   In the quantitative direction typical of soft constraints, there are also other frameworks to model preferences, mostly based on utilities. The most widely used assumes we have some form of independence among variables, such as *mutual preferential independence*. Preferences can then be represented by an additive utility function in deterministic decision making, or *utility independence*, which assures an additive representation for general scenarios [119]. However, this assumption often does not hold in practice since there is usually some interaction among the variables. To account for this, models based on interdependent value additivity have been defined which allows for some interaction between the variables while preserving some decomposability [72]. This notion of independence, also called *generalized additive independence* (GAI), allows for the definition of utility functions which take the form of a sum of utilities over subsets of the variables. GAI decompositions can be represented by a graphical structure, called a GAI net, which models the interaction among variables, and it is similar to the dependency graph of a CP-net or to the junction graph of a Bayesian network [8]. GAI decompositions have been used to provide CP-nets with utility functions, obtaining the so-called UCP networks [22].

Multi-criteria decision making.   In multi-criteria decision making [81, 180], one assumes a set of alternatives which are decomposed into the Cartesian product of domains of some attributes (also called criteria). It is assumed that the agent has enough information to order the domain of each attribute, and the goal is to aggregate this information in order to obtain a preference ordering over the set of alternatives. Under a rather weak assumption (namely *order-separability*) [73], the ordering given on the domain of each attribute can be modeled via a utility function mapping each value in the domain to a real value representing the utility the agent has for that assignment. A very natural way to aggregate the utilities is to use a weighted sum, assuming the agent also provides some weights (usually in [0,1]) reflecting the importance of each attribute. Notice that this can be mapped directly into an equivalent weighted CSP where attributes are represented by variables, and there are only unary constraints mapping values in the domains of the variables to costs obtained by multiplying the weight of the attribute with the opposite of the corresponding utility. Despite an attractive simplicity and low complexity, this approach suffers a major drawback. In fact, using an additive aggregation operator, such as a weighted sum, is equivalent to assuming that all the attributes are independent [136]. In practice, this is not realistic, and therefore non-additive approaches have been investigated. Among them, it has been shown that fuzzy (or non-additive) measures and integrals can be used to aggregate mono-dimensional utility functions [137]. A fuzzy (or Choquet) integral, in fact, can be considered as a sort of very general averaging operator that can represent the notions of importance of an attribute and interaction between attributes.

2.8   CONCLUSIONS

Preference modeling is a very active area of research. It is important to know the properties of various preference modeling frameworks, such as their expressivity, conciseness, and computational complexity since the choice of formalism greatly influences the behaviour of the preference reasoning environment. It would be ideal to have a single environment where users can choose how to model preferences, and where different modeling frameworks can be used even within the same problem, with the aim of allowing a more faithful representation of a real-life preference problem. Much work has been done towards this goal. However, a lot more work is still needed to fully achieve it.

CHAPTER 3

Uncertainty in Preference Reasoning

The formalisms for handling preferences described in the previous chapter support a powerful combination of compactness and expressive power. However, they still require a lot of information from the user. For example, if we represent an agent's preferences with a soft constraint problem with 4 variables, each having 5 domain values, even if we assume only 4 constraints, each involving 2 variables, the user will have to provide 100 preference values (one for each tuple). Thus, it will often be the case that, in the problem specification, the preferences will be provided in an imprecise or vague way, or it will be missing. There are two main approaches to uncertainty in preferences. The first one is to do the best that we can with the available data without further bothering the user. This translates to looking for solutions that are of high quality with respect to the preferences that are known and "robust" with respect to the ones that are missing. The second strategy is to resort to elicitation, that is, to ask the user for the missing preferences. Eliciting preferences takes time and effort, and users may be reluctant to provide their preferences due to privacy concerns or annoyance, especially when confronted with large combinatorial candidate sets. Thus, issues such as the appropriateness of the preference language made available by the system in terms of elicitation, the identification of criteria that allows one to detect when elicitation is no longer necessary and can be stopped, and computational aspects related to elicitation must be taken into account.

We will now give a brief overview of several different results in the literature that exemplify the different approaches. In this chapter we will focus on uncertainty in preferences from a single-agent perspective. Uncertainty on preferences in multi-agent settings will, instead, be considered in the next chapter.

3.1 INTERVAL-BASED PREFERENCES

In this section, we describe an example of an approach that does not require any further elicitation. In soft constraints, each instantiation of the variables in a constraint are associated with a precise preference or cost value. Sometimes it is not possible for a user to know exactly all these values. For example, a user may have a vague idea of the preference value, or may not be willing to reveal his/her preference for privacy reasons. These forms of imprecision have been handled in [90] by extending soft constraints to allow users to state an interval of preference values for each instantiation of the variables of a constraint. This interval can contain a single element (in this case, we have the usual soft constraints), or the whole range of preference values (when there is complete ignorance about the preference value), or it may contain more than one element but a strict subset of the set of preference values. Such problems are called interval-valued soft CSPs (or also IVSCSPs).

Intervals appear to be a particularly suitable way to represent imprecise information about preferences. In fact, linguistic descriptions of degrees of preference (such as "quite high" or "low" or "undesirable") may be more naturally mapped to preference intervals, especially if the preferences are being elicited from different experts, since they may mean somewhat different things by these terms.

As in SCSPs, we assume we have a set of variables, V, with finite domain D. We will now define a generalized notion of soft constraints allowing for preference intervals.

Interval-valued soft constraint. A interval-valued soft constraint is a pair $\langle f, con \rangle$ where con is the subset of variables of V involved in the constraint and f is a function mapping every assignment of the variables in con to an ordered pair of preferences, representing the lower and upper bound of a preference interval.

An interval-valued soft constraint problem (IVSCSP) consists of a set of variables V, with domain D and a set, C, of interval-valued soft constraints. Let us consider the example depicted in Figure 3.1. This figure shows an IVSCSP, that we call P, defined using fuzzy preferences. The IVSCSP P contains three variables X_1, X_2, and X_3, with domain $\{a, b\}$, and five constraints c_1, c_2, c_3, c_4 and c_5. In particular, c_1 is on variable X_1, c_2 is on $\{X_1, X_2\}$, c_3 on X_2, c_4 on $\{X_2, X_3\}$, and c_5 on X_3. The figure shows the definition of each constraint, giving an interval (or a single value) for each variable assignment.

Figure 3.1: An Interval-Valued Soft Constraint Problem.

In an IVSCSP, a complete assignment of values to all the variables can be associated with an interval as well, by combining all the intervals of the relevant tuples in the constraints. For example, if we consider the IVSCSP P of Figure 3.1, the preference interval associated with the complete assignment $s_1 = \langle X_1 = a, X_2 = a, X_3 = a \rangle$ is $[L(s_1), U(s_1)]$, where $L(s_1) = min(1, 0.8, 0.6, 0.8, 0.9) = 0.6$, and $U(s_1) = min(1, 1, 0.95, 0.9, 0.9) = 0.9$. Figure 3.2 shows all the complete assignments (i.e., all the solutions) of P with their preference interval.

When confronted with uncertainty on the value of some parameters, it may be useful to consider the possible ways in which they can be fixed. Each possible way of fixing an uncertain parameter corresponds to a possible outcome and is called a *scenario*. In the context of IVSCSPs, a scenario is a soft constraint problem obtained by choosing one element in each preference interval.

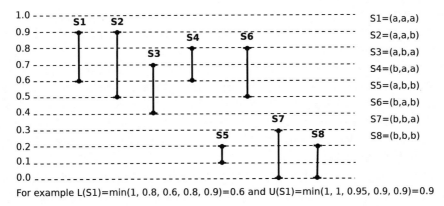

For example L(S1)=min(1, 0.8, 0.6, 0.8, 0.9)=0.6 and U(S1)=min(1, 1, 0.95, 0.9, 0.9)=0.9

Figure 3.2: Solutions of the IVCSP shown in Figure 3.1.

Scenarios can be used to assess the robustness of different solutions. In particular, the two following notions of optimal solutions arise naturally.

Possibly Optimal Solution. An assignment s to all the variables is possibly optimal if it is optimal in some scenario. In other words, there exists a choice of preferences in the intervals such that s is an optimal solution of the resulting SCSP.

Necessarily Optimal Solution. An assignment s to all the variables is necessarily optimal if it is optimal in all scenarios. In other words, no matter how the elements of the preference intervals are fixed, there is no solution with higher preference than s.

If necessarily optimal solutions exists, any algorithm, optimizing over preference and robust to uncertainty, should return one of them. Unfortunately, it is often the case that there is no such solution due to the amount of uncertainty present in the problem. Thus, one may be satisfied with a solution which is optimal in some cases. These definitions can also be weakened to require a given preference level, say α, in, respectively, some or all scenarios. In particular, we say that an assignment is necessarily (resp., possibly) of at least preference α if it has a preference greater or equal than α in all (resp., some) scenarios.

The notions of optimality defined above are very general and may be meaningful in many of the contexts featuring uncertainty. Indeed, we will refer to them again in the following sections of this chapter. In the specific case we are considering, however, where uncertainty is modeled via intervals, ad hoc notions of optimality can be considered. For example, *lower optimal* (resp., *upper optimal*) solutions are assignments that dominate all other assignments with respect to the lower (resp., upper) bound of the associated intervals. If we intersect the sets of lower and upper optimal solutions, we get *interval optimal* solutions, that is, solutions dominating on both bounds. We may require even more, and consider solutions whose lower bound dominates the upper bound of all other solutions. Such solutions are called *interval dominant*.

If we consider our running example, s_1 and s_4 are lower optimal, while s_1 and s_2 are upper optimal. This implies that only s_1 is interval optimal. Moreover, as it can be seen in Figure 3.2, there are no interval dominant solutions.

Interestingly, these interval-based optimality notions can be related to scenarios. For example, interval dominant solutions coincide with the necessarily optimal solutions. Also the other notions can be expressed in terms of scenarios. This means that finding optimal solutions of IVCSPs according to all of the above criteria can, thus, be achieved by solving one or more "standard" SCSPs. This bounds the complexity of solving these kinds of problems with uncertainty to that of solving problems without uncertainty.

Other works have investigated representing uncertainty in preferences via intervals. For example, intervals are used in [151] to deal with unstable costs, which are present in many real-life problems. A typical example is the budget estimate for next year in a company. Such an estimate may be based on data which is not known or not certain, and most of the time such uncertainty is represented as last year's value for that kind of data (which can be seen as the default value), plus some range of possible other values around the default value. Another kind of problem where unstable values may occur is when we want to represent linguistic concepts numerically, such as "more or less", "around", "at least", or "at most". In all these cases, the natural formulation is to have a value and a range around (or above, or below) such a value. A possible approach to deal with this kind of problems is that of *robust optimization* [1], that is, to cast them into uncertain linear programming problems, uncertain conic quadratic and semi-definite optimization problems, or dynamic (multi-stage) problems.

In [151], instead, a framework similar to that of IVSCSPs is considered in which the user can additionally provide an interval of preferences, indicate an element in such an interval which should be considered as the "default" (or most likely) value. In such a setting, optimality notions give priority to optimization with respect to the default values and model different levels of robustness to change in the default values.

Preference intervals have been considered also in the context of multi-criteria decision making [33]. As we have mentioned in the previous chapter, multi-criteria decision making aims at ordering multi-dimensional alternatives. In such a context, mono-dimensional preferences (usually called utilities) are often aggregated using the Choquet integral, which can be regarded as a sort of weighted mean taking into account the importance of every attribute (or criterion). In this context, in order to define preferences over alternatives, the user is required to provide importance and interaction indices among the attributes. The interaction indices usually belong to the interval [-1,+1] and model the fact that two attributes may be complementary (positive index), redundant (negative index), or independent (index equal to 0). However, it is more likely for the user to define intervals of values rather than precise values. Such interval information can be integrated in the scheme of computation of the Choquet integral, by extending its definition to Interval Arithmetic [33].

Other interesting issues regard the intrinsic nature of preferences and its relation with interval-based models. This line of investigation is pursued, for example in [142, 181], where the authors, on the one hand, investigate the necessary and sufficient conditions for which the preference statements of a decision maker can be represented through the comparison of intervals, and, on the other hand, discuss general models through which the comparison of intervals can lead to the establishment of preference relations. The result is a general framework enabling one to clarify the different preference models that can be associated to the comparison of intervals, such as, for example, situations of crisp or continuous hesitation of the decision maker between stating a strict ordering or indifference on two options.

3.2   MISSING PREFERENCES

The increasing use of web services and of multi-agent applications demands the formalization and handling of data that is only partially known when the solution process begins, and that can be added later. In many web applications, data may come from different sources, which may provide their information at different times. Also, in multi-agent settings, data provided by some agents may be hidden for privacy reasons, and only released if needed to find a solution to the problem.

In the previous section, we have seen how soft constraints can be extended via intervals in order to handled imprecise or vague information regarding preferences. We now present a framework that deals with an even stronger assumption: that some preferences are completely missing [88].

In some settings, users may know all the preferences but are willing to reveal only some of them at the beginning. Although some of the preferences can be missing, it could still be feasible to find an optimal solution. If not, the idea is to ask the user to provide some of the missing preferences and to try to solve the new problem.

As in the case of intervals, we have a set of variables V with finite domain D. The first step is to extend the notion of soft constraint.

Incomplete soft constraint. An incomplete soft constraint is a pair $\langle idef, con \rangle$ where con is the subset of the variables in V involved in the constraint and $idef$ is a function mapping every assignment of the variables in con to either a preference or the special value ?, denoting that the preference is unknown. All tuples mapped into ? by $idef$ are called incomplete tuples.

In an incomplete soft constraint, the preference function can either specify the preference value of a tuple by assigning a specific element from the carrier of the c-semiring (see Section 2.2), or leave such preference unspecified. Formally, in the latter case the associated value is ?. We note that a soft constraint is a special case of an incomplete soft constraint where all the tuples have a specified preference.

An incomplete soft constraint problem (ISCSP) consists of a set of variables V with finite domain D and a set of incomplete soft constraints C. Given an ISCSP P, we will denote with $IT(P)$ the set of all incomplete tuples in P.

Let us consider the following scenario as an example of a problem that can be modeled as an incomplete soft constraint problem. A travel agency is planning Alice and Bob's honeymoon. The candidate destinations are the Maldive islands and the Caribbean, and they can decide to go by ship or by plane. To go to Maldives, they have a high preference to go by plane and a low preference to go by ship. For the Caribbean, they have a high preference to go by ship, and they don't give any preference on going there by plane.

Assume we use the fuzzy c-semiring $\langle [0, 1], max, min, 0, 1 \rangle$ (see Section 2.2). We can model this problem by using two variables T (standing for *Transport*) and D (standing for *Destination*) with domains $D(T) = \{p, sh\}$ (p stands for *plane* and *sh* for *ship*) and $D(D) = \{m, c\}$ (m stands for *Maldives*, c for *Caribbean*), and an incomplete soft constraint $\langle idef, con \rangle$ with $con = \{T, D\}$. The only incomplete tuple in this soft constraint is (p, c).

Also, assume that for the season being considered, the Maldives are slightly preferable to the Caribbean. Moreover, Alice and Bob have a high preference for plane as a means of transport, while they don't give any preference to ship. Moreover, as far as accommodations, which can be in a standard room, a suite, or a bungalow, assume that a suite in the Maldives is too expensive while a standard room in the Caribbean is not special enough for a honeymoon. To model this new information, we use a variable A (standing for *Accommodation*) with domain $D(A) = \{r, su, b\}$ (r stands for *room*, *su* for *suite* and *b* for *bungalow*), and three constraints: two unary incomplete soft constraints, $\langle idef1, \{T\} \rangle$, $\langle idef2, \{D\} \rangle$ and a binary incomplete soft constraint $\langle idef3, \{A, D\} \rangle$. The definition of all constraints is shown in Figure 3.3. The set of incomplete tuples of the entire problem is $IT(P) = \{\langle T = sh \rangle, \langle T = p, D = c \rangle, \langle A = su, D = c \rangle, \langle A = b, D = c \rangle, \langle A = r, D = m \rangle, \langle A = su, D = m \rangle\}$.

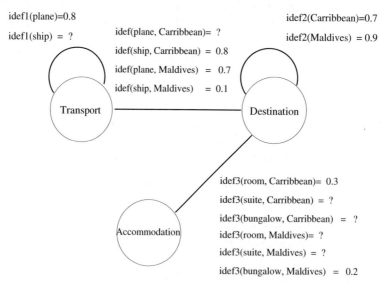

Figure 3.3: An incomplete soft constraint problem.

In ISCSPs, a notion similar to that of a scenario for IVSCSPs is called a *completion*. More specifically, given an ISCSP P, a *completion* of P is an SCSP P' obtained from P by associating a preference with each incomplete tuple in every constraint a preference. A completion may be partial if some preference remains unspecified.

When some of the preferences are missing, it is not clear what the preference induced by the constraints on a complete assignment should be. In the ISCSP framework, given an assignment s, such a preference is obtained by combining the known preferences associated with the projections of the assignment. Such projections are the appropriate sub-tuples in the constraints. The projections which have unspecified preferences are simply ignored. Notice that ignoring such tuples in the global preference computation is the same as assuming they have a preference equal to 1 since 1 is the neutral element for constraint combination. This can been seen as an optimistic approach to uncertainty. For example, if we consider the two assignments $s_1 = \langle T = p, D = m, A = b \rangle$ and $s_2 = \langle T = p, D = m, A = su \rangle$ for our running example shown in Figure 3.3, we have that $pref(P, s_1) = min(0.8, 0.7, 0.9, 0.2) = 0.2$, while $pref(P, s_2) = min(0.8, 0.7, 0.9) = 0.7$. However, while the preference of s_1 is fixed, since none of its projections is incomplete, the preference of s_2 may become lower than 0.7, depending on the preference of the incomplete tuple (su, m). As shown by the example, the presence of incompleteness partitions the set of assignments into two sets: those which have a certain preference which is independent of how incompleteness is resolved, and those whose preference is only an upper bound, in the sense that it can be lowered in some completions.

In SCSPs, an assignment is an *optimal solution* if its global preference is undominated. This notion can be generalized to the incomplete setting by using the same notions that have been introduced regarding intervals. In particular, when some preferences are unknown, we will speak of necessarily and possibly optimal solutions, that is, assignments which are undominated in all (resp., some) completions.

In what follows, given an ISCSP P, we will denote with $NOS(P)$ (resp., $POS(P)$) the set of necessarily (resp., possibly) optimal solutions of P. Notice that, while $POS(P)$ is never empty, in general $NOS(P)$ may be empty. In particular, $NOS(P)$ is empty whenever the available preferences are not sufficient to establish the emergence of an assignment as unconditionally optimal.

In the ISCSP P of Figure 3.3, we can easily see that $NOS(P) = \emptyset$ since, given any assignment, it is possible to construct a completion of P in which it is not optimal. On the other hand, $POS(P)$ contains all assignments not including tuple $\langle T = sh, D = m \rangle$.

In the context of missing preferences, completions play a fundamental role. Among the many ways in which the missing preferences can be completed, let us focus on the completions obtained replacing ? everywhere with the worst possible (respectively, the best possible) preference. These SCSPs are called the worst and best completions, respectively. They are of particular interest since, by computing the preference associated to their optimal solutions, it is possible to understand if the information in the ISCSP is sufficient for the existence of a necessarily optimal solution. In fact, it has been shown that in most cases a necessary and sufficient condition for $NOS(P)$ not to be empty is that the optimal preferences of the worst and best completion coincide [88]. This is an important result which has been used as a stopping condition in a procedure for solving ISCSPs that interleaves search and elicitation of the missing preferences.

3.2.1 INTERLEAVING COMPLETE SEARCH AND ELICITATION

The first general solution strategy that can be applied when there are missing preferences is to combine a branch-and-bound search (B&B) (see Section 2.1) with elicitation steps in which the user is asked to provide some type of missing information. We recall here that B&B proceeds by considering the variables in some order, by choosing a value for each variable in the order, and by computing, using some heuristics, an upper bound on the global preference of any completion of the current partial assignment. B&B also stores the highest preference (assuming the goal is to maximize) of any complete assignment found so far. If at any step the upper bound is lower than the preference of the current best solution, the search backtracks.

When some of the preferences are missing, as in ISCSPs, the agent may be asked for some preferences or other information regarding the missing preferences in order to learn the true preference of a partial or complete assignment, or in order to choose the next value for some variable. Preferences can be elicited after each run of B&B or during a B&B run while preserving the correctness of the approach. For example, we can elicit preferences at the end of every complete branch (that is, regarding preferences of every complete assignment considered by B&B), or at every node in the search tree (thus considering every partial assignment). Moreover, when choosing the value for the next variable to be assigned, we can ask the user (who knows the missing preferences) for help. Finally, rather than eliciting all the missing preferences in the possibly optimal solution, or the complete or partial assignment under consideration, we can elicit just some of the missing preferences.

For example, with incomplete fuzzy constraint problems (IFCSPs), eliciting just the worst preference among the missing ones is sufficient since only the worst value is important for the computation of the overall preference value. In other words, the user is provided with the list of missing preferences and is asked to reveal one which is minimal among them. By contrast, with incomplete weighted constraint problems (IWCSPs), we need to elicit all the missing preference values to decide whether the current assignment is better than the best one found so far.

It is thus possible to design a solving method parametrized by the values of three features which we call *who, what* and *when*. There are four possible choices for *who* (the next value to assign to a variable). This choice can be made by the algorithm in decreasing preference order in the best completion (denoted with *dp*), or in the worst completion (denoted with *dpi*). Otherwise, the choice can be made by the user by looking only at the preferences over the domain values (denoted with *lu*), or by also taking into account the constraints involving that variable (denoted by *su*). If we work with IFCSPs, there are two possibilities for *what* is elicited: *all* the preferences or only the *worst* one. Finally, there are three options for *when* to elicit: at the end of the search *tree*, at the end of every *branch*, or at every *node*. If *when = tree*, elicitation takes place only when the search is completed. This means that the B&B search can be performed more than once. In contrast, if *when = branch* or *when = node*, the B&B search is performed only once and the elicitation is done either at every node of the search tree or at every leaf.

By choosing a value for each of the three parameters above in a consistent way, we obtain, for IFCSP, a total of 16 different algorithms, as summarized in Figure 3.4.

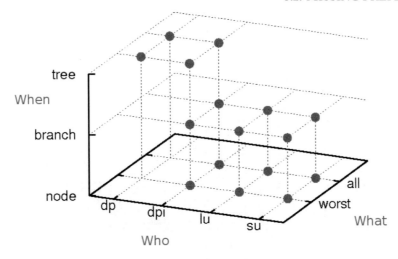

Figure 3.4: The algorithms for IFCSPs.

All possible instances of this schema, obtained by selecting different elicitation strategies, have been tested on randomly generated soft constraint problems (fuzzy and weighted), by varying the number of variables, the tightness and density of constraints, as well as the percentage of missing preferences. The quality measures computed are the percentage of elicited preferences, that is, the missing values that the user has to provide to the system because they are requested by the algorithm, and the percentage of preference values the user may have to compute to be able to respond to the elicitation requests (the *user's effort*). Figure 3.5 summarizes the performance of one of the best algorithms for the fuzzy case when incompleteness varies, the constraint density of the problem is 50% and the constraints forbid around 10% of the tuples. Such an algorithm is obtained when the user is not involved in the selection of the next variable that should be instantiated, and only the worst preference is elicited at the end of every branch. As it can be seen, even when all the preferences are missing, only slightly more than 10% of them need to be elicited (that is, explicitly communicated by the user to the system), and the user must consider around 50% of them.

3.2.2   INTERLEAVING LOCAL SEARCH AND ELICITATION

Incomplete search methods, such as local search methods, can also be adapted to combine solving with preference elicitation [89]. The local search approach for ISCSPs starts by randomly generating an assignment of all the variables. To assess the quality of such an assignment, its preference is computed, asking the user to reveal all or part of the preferences involved in the assignment. Then, at each step, when a variable is chosen, its local preference is computed by setting all the missing preferences to the best possible preference value. To choose the new value for the selected variable, the preferences of the assignments obtained by choosing the other values for this variable are computed.

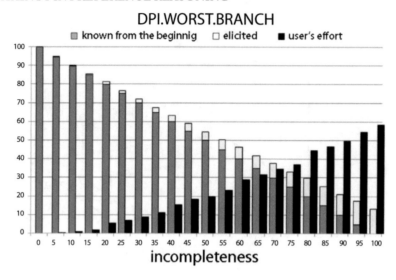

Figure 3.5: Solving incomplete Fuzzy SCSPs by interleaving B&B search and elicitation.

Since some preference values may be missing, in computing the preference of a new assignment only the preferences which are known at the current point are considered. The value which is associated with the best new assignment is then chosen, using the number of incomplete tuples as a tie breaker. Thus, the procedure moves to a new assignment which is better than the current one and has the fewest missing preferences.

Since the new assignment could have incomplete tuples, the user is asked to reveal enough of this data to compute the actual preference. In the case of fuzzy preferences, for example, it is enough to ask the user to reveal only the worst preference among the missing ones, if it is less than the worst known preference.

As in many classical local search algorithms, to avoid stagnation in local minima, tabu search and random moves are used. During search, the algorithm maintains the best solution found so far, which is returned when the maximum number of allowed steps is exceeded.

When compared against the best complete algorithm, this local search approach in general asks the user for a few more preferences but its scaling behavior is much better. Also, the preference of the solution found is very close to that of the (necessarily optimal) solution found by the complete approach.

3.3   CONCLUSIONS

In many real-life settings, the specification of preferences is unavoidably characterized by vagueness and imprecision. This may be due to the large amount of information involved, privacy concerns, or communication costs. Compact preference models are appealing in this respect since they often suggest a structured way to embed such uncertainty into the model. We have seen that uncertainty about the preferences can be total or partial. Soft constraint problems allow modeling of both cases, for example, by using intervals of preferences. We also discussed two main ways to deal with uncertainty, depending on whether elicitation is an option or not. If elicitation is not an option, compact preference models can help in computing the best among the solutions that are robust with respect to the missing information. If, instead, elicitation can be performed, it can be guided by the compact structure of the problem and by the solution technique adopted. There are many other types of uncertainty that concern preferences, such as the presence of uncontrollable variables [150]. Also, it is important to consider uncertainty in conditional preferences as modeled by CP-nets (see Chapter 2). Thus, there is still much work to be done in this respect. Since handling uncertainty is often a requirement for the actual deployment of intelligent systems dealing with preferences, this interesting line of research will continue for several years to come.

CHAPTER 4

Aggregating Preferences

An important aspect of reasoning about preferences is preference aggregation. In multiagent systems, we often need to combine the preferences of several agents. For instance, SCATS (the Sydney Coordinated Adaptive Traffic System) is a complex distributed multiagent system used to control the traffic lights in Sydney and 140 other cities around the world [128]. The system is distributed as each intersection is controlled by a separate kerbside computer. Based on the current traffic demands, each kerbside computer in a SCATS system has a preferred plan for the cycle time for the traffic lights at its intersection, as well as the ratio of this time given to the different approaches. To ensure traffic flows smoothly, a coordination mechanism is needed to choose between these different preferred plans.

A simple, effective and well understood mechanism for aggregating together preferences of multiple agents is to apply a voting rule. Each agent expresses a preference ordering over the set of candidates, and an election is held to compute the winner. Going back to the SCATS example, each kerbside computer casts a vote for its preferred plan. A new plan is then adopted when it receives a sufficient threshold of votes. Social choice studies the properties of methods for aggregating individual preferences. Social choice has thus been of considerable interest within multiagent systems. However, as argued before, multiagent systems raise some fresh issues for social choice, mostly due to computational features of the problem [35, 36, 68, 123]. Ideas from voting have also found application in a number of other computational areas like rank aggregation [60] and recommender systems [146]. In all these areas, the number of candidates running in the election can be large. This feature can give rise to significant new computational issues.

4.1 VOTING

Mankind has been voting for thousands of years. Whilst Athenian citizens could all vote in the Assembly around 500BC, it is believed that democracies existed in Mesopotamia and India at least one century before this. More recently, starting with the work of people like the marquis de Condorcet in the 18th century, attention has focused on understanding the properties of different voting methods, as well as on inventing new methods with good properties.

Perhaps the simplest voting rule is plurality voting. Each voter identifies whichever of the candidates they prefer most, and the candidate with the most votes wins. With two candidates, this is also called majority voting. There is little more to say about elections between two candidates. There are strong theoretical reasons to choose majority voting when there are just two candidates. Many other voting rules degenerate to majority voting when we have just two candidates to choose between. In addition, May's theorem proves that majority voting is the only rule over two candidates that satisfies some rather weak but natural "fairness" properties [132]. We shall introduce these properties formally in a short time, but briefly, they are anonymity (each voter is treated the same), neutrality (each candidate is treated the same) and monotonicity (increasing your preference for an candidate does not hurt this candidate).

Unfortunately, when there are more than two candidates in the election, voting is more problematic. Indeed, as we shall describe shortly, there are axiomatic issues which suggest that all voting rules have drawbacks. For example, one problem is the existence of Condorcet cycles. Suppose there are three voters choosing their favourite Musketeer, and the preferences of the three voters are as follows:

Voter 1:	Athos	$\succ$	Porthos	$\succ$	Aramis
Voter 2:	Porthos	$\succ$	Aramis	$\succ$	Athos
Voter 3:	Aramis	$\succ$	Athos	$\succ$	Porthos

This can be read as Voter 1 prefers Athos to Porthos, and Porthos to Aramis, etc. In this situation, a majority of voters (two out of three) prefer Athos to Porthos, and Porthos to Aramis. However, a majority of voters prefer Aramis to Athos. There is no Musketeer that a majority of voters prefer to all others. The collective preferences are thus cyclic and not transitive. The source of this paradox is that the different majorities are made up of different voters. It leaves us with the problem of how we break such cycles. We must declare one winner even though each candidate is in an exactly symmetric situation.

4.1.1   VOTING RULES

Many different voting rules have been proposed over the years. Indeed, new rules continue to be proposed to deal with issues like Condorcet cycles. We will formalise voting rules as follows. A *profile* is a sequence of n total orders over m candidates. A *voting rule* is a function mapping a profile onto one candidate, the *winner*. We let $N(i, j)$ be the number of voters preferring i to j, and $beats(i, j)$ to be 1 if and only if $N(i, j) > \frac{n}{2}$ and 0 otherwise. We shall consider some of the most common voting rules on profiles (see, for instance, [29, 44]).

Plurality: The candidate ranked in first place by the largest number of voters wins. For example, plurality is used to elect Members of the British Parliament. When there are only two candidates, plurality is the same as majority voting.

Scoring rules: Suppose there are m candidates, $(w_1, \ldots, w_m)$ is a vector of weights where the ith candidate in a vote scores w_i, then the winner is the candidate with highest total score over all the votes. The *plurality* rule has the weight vector $(1, 0, \ldots, 0)$, the *veto* rule has the vector $(1, 1, \ldots, 1, 0)$, whilst the *Borda* rule has the vector $(m - 1, m - 2, \ldots, 0)$. The Borda rule was used to elect members of the French Academy of Sciences until this was over-turned by Napoleon. It is still used today to elect minority members of the National Assembly of Slovenia, as well as in various organizations including Robocup. Although Jean-Charles de

Borda proposed the method in 1770, Ramon Llull may have invented it as early as the 13th century.

Cup: The winner is the result of a series of pairwise majority elections (or matches) between candidates. The rule requires the sequence of matches to be given. This is called the *agenda*. Each root is labelled with a candidate, and the node of a parent is labelled with the winner of the pairwise majority election between the two children. The winner is the label of the root.

Copeland: The candidate with the highest Copeland score wins. The Copeland score of candidate i is $\sum_{j\neq i} beats(i,j) - beats(j,i)$. The Copeland winner is the candidate that wins the most pairwise elections. In the second-order Copeland rule, if there is a tie, the winner is the candidate whose defeated competitors have the largest sum of Copeland scores. The Copeland rule is used, for example, by the World Chess Federation to decide the winner of a round robin tournament. A variant of the Copeland rule is also credited to Ramon Llull [66]. With the *Llull* rule, both candidates score a point if they tie in their pairwise election. By comparison, with the Copeland rule, neither candidate gets a point in this situation. A third variant is when each candidate gets a $\frac{1}{2}$ point in the case of a pairwise tie.

Simpson (also known as maximin): The candidate with the highest maximin score wins. The maximin score of candidate i is $\min_{j\neq i} N(i,j)$. Thus, the Simpson winner is the candidate whose worst performance in pairwise elections is best.

Plurality with runoff: If one candidate has a majority, they win. Otherwise, everyone but the two candidates with the most votes are eliminated, and the winner is chosen using the majority rule. A variation of this rule (used in the French National Assembly) is to eliminate all candidates with less than a given threshold of votes and then run a plurality election.

Single Transferable Vote (STV): This rule requires up to $m-1$ rounds. In each round, the candidate with the least number of voters ranking them first is eliminated until one of the remaining candidates has a majority. STV is used in a number of parliamentary elections, as well as by many organizations including the General Synod of the Church of England and the Academy Awards. STV is also known as the Hare or the Hare-Clark rule. Thomas Hare is generally credited with proposing STV voting in 1857, although the idea of transferable votes can be traced further back to Thomas Wright Hill in 1821. In 1896, Andrew Inglis Clark was responsible for the introduction of STV into parliamentary elections in Tasmania where it remains in use today.

Nanson's and Baldwin's rules: These are iterated versions of the Borda rule. In Nanson's rule, we compute the Borda scores and eliminate any candidate with less than half the mean score. We repeat until there is an unique winner. In Baldwin's rule, we compute the Borda scores and eliminate the candidate with the lowest score. We again repeat until there is an unique winner. Just as STV can be seen as an iterated version of plurality voting, these two rules can be seen as iterated versions of the Borda rule. Both rules have been used in real world elections. Nanson's rule, for example, is in use in elections in the University of Adelaide, whilst Baldwin's rule was used by the Dialectic Society of Trinity College, the oldest collegiate society in Australia.

We also consider some other common voting rules, besides plurality, in which voters do not provide a complete ordering over the candidates but either give a set of preferred candidates or provide scores for some or all of the candidates.

Approval: Each voter labels candidates as approved or not. The candidate with the most number of approvals wins. Approval voting is used by many professional organizations including the Mathematical Association of America, the Institute of Management Sciences and the American Statistical Association. The approval rule is often used when multiple candidates need to be elected. In this case, we elect the k candidates with the most approvals.

Range: Each voter assigns a number from a given range to each candidate. The candidate with the highest total wins. Approval voting is an example of range voting where only the integers 0 or 1 are allowed. Range voting is used in many sporting events (e.g., to judge gymnastic competitions).

Cumulative: Voters have a number of points to distribute between candidates. The candidate with the highest total number of points wins. Cumulative voting was used for over 100 year to elect the Illinois House of Representatives. It is often used in corporate governance, and it is in fact mandated by several U.S. states.

In some situations, the required result of an election is not one winner but a set of winners. For instance, we might want to select k people to sit on a committee. Many of these voting rules can be readily adapted to select multiple winners. There are also specific voting systems (technically, they are called *voting correspondences*) for selecting multiple winners. For example, in bloc voting, each voter gives one point to k candidates and the k candidates with the most points are selected. All these voting rules can also be easily modified to work with *weighted* votes. A vote of integer weight k can be viewed as k voters who vote identically.

In most cases, voting rules are computational cheap to apply. Indeed, those used in practice are typically computable in linear or quadratic time in the number of candidates, and almost always linear in the number of voters. Therefore, computing the outcome of an election is usually not computationally difficult. For example, all of the voting rules discussed above are cheap to compute. However, this is not always the case, and there have been rules proposed in the literature which are more computationally complex to apply. For example, the Victorian mathematician Charles Dodgson (better known as Lewis Carroll, the author of "Alice's Adventures in Wonderland" and "Through the Looking-Glass") proposed one such rule in 1876. In Dodgson's rule, the election is won by the candidate who is closest to being a Condorcet winner, the candidate that would win a pairwise election against every other candidate. Distance is measured by the number of exchanges of adjacent preferences in the voters preference orders needed to make the candidate a Condorcet winner. Computing such a winner is NP-hard [12].

4.1.2 PROPERTIES OF VOTING RULES

One way to study different voting rules is to consider the (desirable) properties that they possess [178]. This is referred to as an *axiomatic* approach, and it can be based either on *social welfare* functions that take a profile and return a complete order on candidates or on *social choice* functions that return just a single winner. We can convert the voting rules described above into social welfare functions in straightforward ways. For instance, any scoring rule can be converted into a social welfare function by ordering candidates according to their score, whilst STV can be converted into a social welfare function by ordering candidates according to their elimination order. A large number of simple properties have been put forwards that social choice and social welfare functions might possess.

Anonymity: Each voter is treated the same. In particular, if we change the order of the voters then the result is unchanged.

Neutrality: Each candidate is treated the same. More precisely, if we permute the names of two candidates then the result is just the permutation of the previous result.

Surjectivity: For every possible result, there exists a profile which gives this result.

Pareto efficiency: For a social choice function, if everyone most prefers one candidate then this candidate wins. For a social welfare function, if everyone prefers one candidate to another, then this is the case in the result.

All the voting rules defined above possess these properties of anonymity, neutrality, surjectivity and Pareto efficiency. However, there are other, more complex properties which might be considered desirable that are not possessed by several voting rules.

Monotonicity: Increasing preferences for an candidate does not hurt the candidate. In particular, if we move a candidate up a voter's preference order, then the candidate should not move lower in the result. For example, the plurality voting rule is monotonic. By comparison, plurality with runoff is not monotonic due to the way votes are "transferred" between rounds.

Independence of irrelevant alternatives: The ordering between two candidates only depends on the individual preferences between these two candidates. Changes in preferences for a third, irrelevant candidate have no impact on the result. For example, approval and range voting both satisfy the appropriate form of independence of irrelevant alternatives. By comparison, most of the other voting rules including plurality and Borda are not independent of irrelevant alternatives.

Majority criterion: If a majority of voters most prefer a single candidate then this candidate wins. For example, plurality, Copeland and STV rules satisfy the majority criterion. On the other hand, Borda and range voting do not satisfy the majority criterion.

Condorcet consistency: A *Condorcet winner* is a candidate who beats all others in pairwise elections [139]. Not all elections have a Condorcet winner. For instance, in our previous example of the three Musketeers, there is a Condorcet cycle and thus no Condorcet winner. A voting rule which elects the Condorcet winner when this exists is called *Condorcet consistent*. For example, Copeland, Nanson's and Baldwin's voting rules are all Condorcet consistent. By comparison, many of the other voting rules including plurality and Borda are not Condorcet consistent.

Participation: Voters have an incentive to vote. In particular, if we add a voter for whom one candidate is preferred to another, then the winner should not change from the first to the second candidate. For example, plurality, Borda and approval voting rules satisfy participation. On the other hand, no Condorcet consistent method satisfies the participation criterion.

This list of properties is not exhaustive (though it does represent some of the most important properties). There are many other properties which have been proposed and studied. For example, a voting rule satisfies the *Condorcet loser criterion* if a candidate loses when they lose to every other candidate in pairwise elections. As a second example, a voting rule satisfies *reversal symmetry* if the winner loses when all preferences are inverted. Many of these properties are uncontroversial (e.g., anonymity is consistent with "one person, one vote", whilst participation encourages voters to vote). However, the property of independence of irrelevant alternatives has received considerable debate. For example, many authorities argue that it ignores different intensity of preferences whilst others argue that people's preferences often change when new alternatives are introduced.

4.1.3   FAIRNESS AND MANIPULATION

A seminal result in social choice theory is Arrow's famous impossibility result that many of these properties are incompatible with the voting rule being "fair" [7]. Fairness is defined in terms of the absence of a dictator.

Dictatorial: One voter alone decides the outcome. More precisely, there exists one voter whose vote is the result.

We clearly do not want just one voter's preferences to matter. It seems reasonable to expect that the final result will depend collectively on everyone's preferences. Unfortunately, Arrow's theorem proves that this is not possible under weak assumptions like three or more candidates to choose from:

Theorem 4.1   Arrow's theorem.   If there are three or more candidates, and the social welfare function is Pareto efficient and independent of irrelevant alternatives, then it is dictatorial [160].

It follows from Arrow's theorem that voting rules are vulnerable to "strategic voting". From our perspective, this means voters may have an incentive to mis-report their preferences. The notion of strategic voting is captured in the following definition.

Strategy proof: It is best for each voter to order outcomes as they prefer. More precisely, a voter cannot improve the outcome by mis-reporting their preferences.

A voting rule that is strategy proof is also called non-manipulable. The Gibbard-Satterthwaite theorem [93, 172] is closely related to Arrow's result and states:

Theorem 4.2 Gibbard-Sattertwhaite theorem.   If there are three or more candidates, and the social choice function is strategy proof and surjective, then it is dictatorial [160].

Voters thus may be able to profit by voting tactically. Consider again our three Musketeer example and plurality voting where ties are broken alphabetically. Suppose a fourth voter most prefers Athos. Then, from the four votes, Athos will have two votes, and Aramis and Porthos a single vote. Consider a fifth voter who prefers Porthos to Aramis and Aramis to Athos. If this fifth voter votes sincerely for their most preferred choice, that is, Porthos, then Athos wins. However, if this fifth voter votes strategically for their second most preferred choice, that is, Aramis, then the result is more desirable for them as Aramis wins according to the tie-breaking rule.

Manipulation is especially problematic in multiagent systems as agents may have significant computational resources at their disposal to invest in manipulating the outcome. Unfortunately, the assumptions of the Gibbard-Sattertwhaite theorem are relatively modest, so it seems likely that we will fall within their scope in most situations.

4.1.4 SINGLE-PEAKED PREFERENCES

One possible escape from the Gibbard-Sattertwhaite theorem is to give up the notion of a universal domain in which all preferences are possible, but to consider a restricted class of preferences. For example, one such restriction is to *single-peaked* preferences [96]. Preferences are single-peaked if we can put an order on the candidates, every voter has a most preferred candidate and likes a candidate less as we move away from their preferred candidate. Single-peaked preferences are natural when we are considering features like price (e.g., "I have a preferred budget to spend on an apartment, and the more I have to under-spend or over-spend, the more I will dislike the choice"). We can sort the most preferred candidates (and the corresponding voters) according to this order. With single-peaked preferences, the median voter is the voter whose most preferred candidate is the median in this order amongst the voters. When preferences are single-peaked, the median voting rule (which elects the candidate most preferred by the median voter) is strategy proof, surjective and not dictatorial. With the median voting rule, it is in the best interests of every voter to announce their most preferred candidate. If they vote for any other candidate, they either do not change the result or make it worse for themselves. Unfortunately, preferences are often multi-dimensional (e.g., "My preference for an apartment depends on price, size and location") and are not single-peaked.

4.2 COMPUTATIONAL ASPECTS OF MANIPULATION AND CONTROL

Another appealing escape from the Gibbard-Satterthwaite theorem was proposed by Bartholdi, Tovey and Trick in an influential paper [11]. Perhaps it is computationally so difficult to find a successful manipulation that voters have little option but to report their true preferences? To illustrate this idea, they demonstrated that the second-order Copeland rule is NP-hard to manipulate. Shortly after, Bartholdi and Orlin proved that the more well known Single Transferable Voting (STV) rule is NP-hard to manipulate [10]. A whole subfield of social choice has since grown from this proposal, proving that various voting rules are NP-hard to manipulate under different assumptions (e.g., [44]). Computational complexity results of this kind typically vary along four different dimensions.

Weighted or unweighted votes: Are the votes weighted or unweighted? Weighted votes are used in a number of real-world settings like shareholder meetings and elected assemblies. Weights are also useful in multi-agent systems where we have different types of agents. Weights are interesting from a computational perspective for at least two reasons. First, weights can increase computational complexity. For example, computing a manipulation for the veto rule is polynomial with unweighted votes but NP-hard with weighted votes [44]. Second, the weighted case informs us about the unweighted case when we have probabilistic information about the votes. For instance, if it is NP-hard to compute if an election can be manipulated with weighted votes, then it is NP-hard to compute the probability of a candidate winning when there is uncertainty about how the unweighted votes have been cast [40].

Bounded or unbounded number of candidates: Do we have a fixed number of candidates? Or is the number of candidates allowed to grow? For example, with unweighted votes, computing a manipulation of the STV rule is polynomial if we bound the number of candidates and only becomes NP-hard when the number of candidates is allowed to grow with the problem size [10]. Indeed, with unweighted votes and a bounded number of candidates, it is polynomial to compute how to manipulate most voting rules [44]. On the other hand, with weighted votes, it is NP-hard to compute how to manipulate many voting rules with only a small number of candidates (e.g., with Borda, it is NP-hard with 3 or more candidates [44]).

One manipulator or a coalition of manipulators: Is a single voter trying to manipulate the results or is a coalition of voters acting together? A single voter is unlikely to be able to change the outcome of many elections. A coalition, on the other hand, may be able to manipulate the result. With some rules, like STV, it is NP-hard to compute how a single voter needs to vote to manipulate the result or to prove that manipulation by this single voter is impossible [10]. With other rules, it may require a coalition of voters of some fixed size or larger before manipulation becomes NP-hard to compute (e.g., with Borda, it is polynomial for a single voter to compute how to manipulate the result but NP-hard for two or more voters). Finally, there are other voting rules where it is NP-hard to compute a manipulation only when the number of manipulators is not bounded (e.g., plurality with runoff with weighted votes [44]).

Complete or incomplete information: Many results assume that the manipulator(s) have complete information about the other voters' votes. Or course, we may not know precisely how other voters will vote in practice. However, there are several reasons why the case of complete information is interesting. First, if we can show that it is computationally intractable to compute a manipulation of the election with complete information, then it is also intractable when we have incomplete information. Second, the complete information case informs us about the case when we have uncertainty. For instance, if it is computationally intractable for a coalition to compute how to manipulate an election with complete information, then it is also intractable for an individual to compute how to manipulate an election when we have only probabilistic information about the votes [44].

Constructive or destructive manipulation: Is the manipulator trying to make one particular candidate win (constructive manipulation) or prevent one particular candidate from winning (destructive manipulation)? Destructive manipulation is easier to compute than constructive manipulation. For instance, constructive manipulation of the veto or the Copeland rules by a coalition of voters with weighted votes is NP-hard but destructive manipulation is polynomial [44]. However, there are also rules where both destructive and constructive manipulation are in the same complexity class (e.g., both constructive and destructive manipulation of plurality are polynomial to compute, whilst both constructive and destructive manipulation of plurality with runoff for weighted votes are both NP-hard [44]).

In Figure 4.1, we give a representative selection of results about the complexity of manipulating voting (see [44] for references to many of these results).

4.2.1 MANIPULATION ALGORITHMS

Despite the intractability in the worst case of solving NP-hard problems, several algorithms have been proposed to compute manipulations. These algorithms can be polynomial (where computing manipulations is polynomial) or exponential (where computing manipulations is NP-hard). The polynomial algorithms typically exploit two properties of many manipulation problems [44]. First, for some but not all voting rules, if there is a manipulation, then there exists a manipulation in which every voter in the manipulating coalition votes identically. Second, for many voting rules it is sufficient for each voter to vote greedily for the candidate they want to win or against the candidate they want to lose. For example, a polynomial algorithm to manipulate constructively the plurality rule is for each member of the coalition to vote for the candidate that they wish to win. As a second example, a polynomial algorithm to manipulate destructively the veto rule is for each member of the coalition to veto the candidate that they wish to lose.

Algorithms have also been proposed to compute manipulations for voting rules like STV where computing a manipulation is NP-hard. For instance, Coleman and Teague give a simple enumerative method for a coalition of k unweighted voters to compute a manipulation of the STV rule which runs in $O(m!(n + mk))$ time where n is the number of voters and m is the number of candidates [39].

For a single manipulator, Conitzer, Sandholm and Lang propose an $O(n1.62^m)$ time algorithm that, given an incomplete profile, computes the set of candidates that can still win a STV election

# candidates # manipulators	unweighted votes, constructive manipulation 1	 ≥ 2	weighted votes, constructive 2	 3	 ≥4	 destructive 2	 3	 ≥4
plurality	P	P	P	P	P	P	P	P
cup	P	P	P	P	P	P	P	P
Borda	P	NP-c	P	NP-c	NP-c	P	P	P
veto	P	P	P	NP-c	NP-c	P	P	P
STV	NP-c	NP-c	P	NP-c	NP-c	P	NP-c	NP-c
plurality with runoff	P	P	P	NP-c	NP-c	P	NP-c	NP-c
maximin	P	NP-c	P	P	NP-c	P	P	P
Nanson	NP-c	NP-c	P	P	NP-c	P	P	NP-c
Baldwin	NP-c	NP-c	P	NP-c	NP-c	P	NP-c	NP-c

Figure 4.1: Computational complexity of deciding if various voting rules can be manipulated by a small number of voters (unweighted votes) or a coalition of voters (weighted votes). P means that the problem is polynomial, NP-c that the problem is NP-complete. For example, constructive manipulation of the veto rule is polynomial for unweighted votes or for weighted votes with 2 candidates, but NP-complete for weighted votes and 3 or more candidates. On the other hand, destructive manipulation of the veto rule is polynomial for weighted votes with 2 or more candidates.

[44]. Walsh has modified this algorithm to compute a manipulation that makes a particular candidate win [189]. The two problems are closely related since a candidate can still win if and only if the manipulator can make this candidate win. The algorithm uses a simple recursion in which we lazily construct the manipulator's vote.

For any of the voting rules introduced above, we can easily make any candidate win provided we have enough manipulators. We can therefore consider manipulation as an optimization problem where we try to minimizing the number of manipulators required. One option is to use approximation methods to tackle such optimization problems. For example, Zuckerman, Procaccia and Rosenschein propose a simple approximation method to compute manipulations of the Borda rule [196]. Their method constructs the vote of each manipulator in turn. The candidate that the manipulators wish to win is put in first place, and the remaining candidates are placed in the manipulator's vote in reverse order of their current Borda scores. The method continues constructing manipulating votes until the desired candidate wins. A rather long and intricate argument shows that this simple greedy method constructs a manipulation which uses no more than one additional manipulator than optimal.

4.2.2   TIE-BREAKING

One aspect of voting that we have not discussed in great detail yet is what happens when the vote is tied. As we saw in the Musketeer example, ties can be problematic. Indeed, ties can have a significant impact on the complexity of computing a manipulation. There are several common mechanisms to deal with ties. In the UK, when an election is tied, the returning officer will choose between the candidates using a random method like lots or a coin toss. The local council elections in Great Yarmouth in May 2010 were decided by one candidate drawing a seven from a deck of cards, whilst the other tied candidate drew a three. A typical assumption made in the literature is that we have an odd number of voters as this may make ties impossible. However, other assumptions are also made. For example, we might assume that ties are broken in favour of the manipulator. Suppose the manipulator can make their preferred candidate win assuming ties are broken in their favour but ties are in fact broken at random. Then we can conclude that the manipulator can increase the chance of getting their preferred result. Tie-breaking can even introduce computational complexity into manipulation. For example, computing how to manipulate the Copeland rule with weighted votes is polynomial if ties are scored with 1 but NP-hard if they are scored with 0 [67].

4.2.3   CONTROL

Strategic voting is just one of several ways that the result of an election can be manipulated. For example, the chair person running the election may be able to manipulate the result by controlling how the election is run [13]. A number of different types of control have been studied:

Candidate addition/deletion: Can the chair add (some given number of) spoiler candidates to make a particular candidate win (or lose)? By deleting (some given number of other) candidates, can the chair make a particular candidate win (or lose)?

Voter addition/deletion: Can the chair add (some given number of) new voters with given preferences in order to make a particular candidate win (or lose)? By deleting (some given number of other) voters, can the chair make a particular candidate win (or lose)?

Candidate partitioning: Can the chair partition the candidates into two sets so that a particular candidate wins (or loses) when all the candidates in the first set go into an election, and the survivors go into an election against the second set?

Run off partitioning: Can the chair partition the candidates into two sets so that a particular candidate wins (or loses) when all the candidates in each set go into an election against each other, and the survivors of both elections go forwards into a final election?

Voter partitioning: Can the chair partition the voters into two sets so that a particular candidate wins (or loses) when we run an election with each set of voters, and the survivors of these two elections go forwards into a final election in which all voters vote?

The chair can even try to manipulate the result by trying multiple methods of attack [64]. For example, the chair might simultaneously remove some voters, add other new voters, and introduce a spoiler candidate. Loosely speaking, we say that a voting rule is *immune* to a particular type of control if the chair cannot change the result. For example, with approval voting, adding spoiler candidates cannot stop a candidate being approved. On the other hand, a voting system is said to be *(computationally) vulnerable* to control if it is not immune to control and computing how the chair can control the election takes polynomial time. For instance, with plurality voting, the chair can easily calculate how many voters need to be deleted to ensure the current winner loses. The chair only needs to delete those voters who put the current winner in first place. Finally, a voting system is said to be *resistant* to control if it is not immune to control but computing what the chair needs to do is computationally intractable (i.e., at least NP-hard). For example, with plurality voting, it is NP-hard for the chair to compute which spoiler candidates to add to ensure a particular candidate wins. Not surprisingly, some of these types of control are related. For instance, a voting system is immune to constructive control by deleting candidates if and only if it is immune to destructive control by adding candidates.

Finally, other types of manipulation have also been considered. For example, a voter may try to change the result by means of bribery [63]. In the simplest setting, we can consider the least number of voters we need to bribe to make a given candidate win. However, other more complex settings have been considered. Each voter may be willing to change their vote to whatever we choose, but only if we can meet a given price. Voters may also have different prices depending on how we want to change their vote. For example, with unweighted votes, both these forms of bribery are polynomial to compute for plurality but NP-hard for approval [63]. Another type of control is when the chair changes the running of the election. For instance, in a cup election, the chair may be able to manipulate the result by re-ordering the agenda [153, 169, 190]. This is closely related to the problem of finding a good (or in some sense optimal) ranking of players in a knock out tournament [175].

4.2.4 HYBRID RULES

New voting rules have been proposed that are designed specifically to be computationally hard to manipulate or control. One general construction is to "hybridize" two or more existing voting rules. For example, we might add one elimination pre-round to the election in which candidates are paired off and only the most preferred goes through [42]. This generates a new voting rule that is often computationally hard to manipulate. In fact, the problem of computing a manipulation can now move to complexity classes higher than NP. For instance, adding such a pre-round to plurality makes computing a manipulation NP-hard, #P-hard, or PSPACE-hard, depending on whether the schedule of the pre-round is determined before the votes are collected, after the votes are collected, or the scheduling and the vote collecting are interleaved, respectively [42]. Such hybrid voting rules also inherit some (but not all) of the properties of the voting rule from which they are constructed. For example, if the initial rule is Condorcet consistent, then adding a pre-round preserves Condorcet consistency.

Other types of voting rules can be hybridized. For example, we can construct a hybrid of the Borda and Simpson rules in which we run two rounds of Borda, eliminating the lowest scoring candidate each time, and then apply the Simpson rule to the remaining candidates. Such hybrids are often resistant to manipulation. For example, many hybrids of STV and of Borda are NP-hard to manipulate [62]. Another way to hybridize two or more voting rules is to use some aspect of the particular election (the preferences of the voters, or the names of the candidates) to pick which voting rule is used to compute the winner. For example, suppose we have a list of k different voting rules. If the candidate names (viewed as natural numbers) are congruent, modulo k, to i then we use the ith voting rule. Such a form of hybridization gives elections which are often computationally difficult to control [101].

4.2.5   MANIPULATION ON AVERAGE

The fact that a particular voting rule is NP-hard to manipulate or to control only indicates that the problem is computationally intractable in the worst case. Assuming P≠NP, this means that there are pathological instances of the problem which will take exponential time to solve. It does not tell us whether the computational problem is hard on average or in practice. Several recent theoretical results suggest that elections are often easy to manipulate on average. For example, Procaccia and Rosenschein proved that for most scoring rules and a wide variety of distributions over votes, when the size of the coalition is $o(\sqrt{n})$, the probability that they can change the result tends to 0, and when it is $\omega(\sqrt{n})$, the probability tends to 1 [158]. They also gave a simple greedy procedure that will find a manipulation of a scoring rule in polynomial time with a probability of failure that is an inverse polynomial in n [159].

As a second example, Xia and Conitzer have shown that for a large class of voting rules including STV, as the number of voters grows, either the probability that a coalition can manipulate the result is very small (as the coalition is too small), or the probability that they can easily manipulate the result to make any alternative win is very large [193]. They left open only a small interval in the size for the coalition for which the coalition is large enough to manipulate but not obviously large enough to manipulate the result easily.

Other theoretical results suggest that manipulation may be computationally easy since a random manipulation has a high probability of succeeding. For instance, Friedgut, Kalai and Nisan proved that if the voting rule is neutral and far from dictatorial and there are 3 candidates, then there exists a voter for whom a random manipulation succeeds with probability $\Omega(\frac{1}{n})$ [79]. Starting from different assumptions, Xia and Conitzer showed that a random manipulation would succeed with probability $\Omega(\frac{1}{n})$ for 3 or more candidates for STV, for 4 or more candidates for any scoring rule, and for 5 or more candidates for Copeland [194].

4.2.6   MANIPULATION IN PRACTICE

Empirical studies have also considered the computational difficult of computing manipulations in practice. For example, Coleman and Teague experimentally demonstrated that only relatively small coalitions are needed to change the elimination order of the STV rule in most cases [39]. On the other hand, they observed that most uniform and random elections are not trivially manipulable using a simple greedy heuristic. More complex methods appear able to compute manipulations in many cases relatively quickly. For instance, Walsh showed that manipulations of the STV rule can be computed for a wide range of random and non-random voting distributions in a reasonable amount of time even with hundreds of candidates [189].

Walsh also empirically studied the computational cost of manipulating the veto rule by a coalition of voters whose votes are weighted [188]. He showed that it is usually easy to find manipulations of the veto rule or prove that none exist for many independent and identically distributed votes, even when the coalition is critical in size. He was able to identify a situation in which manipulation is computationally hard. This is when votes are highly correlated and the election is "hung". However, even a single uncorrelated voter is then enough to make manipulation computationally easy again. These average case, approximation and empirical results suggest that we need to treat computational complexity results about the NP-hardness of manipulation with some care. Voters may still be able to compute a manipulation successfully using rather simple and direct methods.

4.2.7   PARAMETERIZED COMPLEXITY

Another approach to gain insight into the computational complexity of manipulation and control comes from parameterized complexity. Traditional worst case analysis focuses on polynomial problems (for which the running time of the best algorithm is of the form $O(n^c)$ where n is the size of the input and c is some constant) and those problems which are NP-hard (for which the running time of the best known algorithms is not polynomial). This is essentially a one-dimensional view of complexity in terms of the single parameter, the problem size n. Parameterized complexity takes a more refined two-dimensional view of complexity. We try to identify another parameter of the problem such that the computational complexity is restricted to the size of this parameter. More precisely, we say that a problem is *fixed-parameter tractable* (FPT) if it can be solved in $O(f(k)n^c)$ time where f is *any* computable function, k is this new parameter, c is again a constant, and n remains the size of the input. If k is small and bounded in size, then the time to solve the problem remains essentially polynomial.

Not all problems are known to be fixed-parameter tractable. Downey and Fellows have proposed, therefore, a hierarchy of fixed-parameter *intractable* problem classes $W[t]$, for $t \geq 1$ [54]. $W[t]$ is characterized by the maximum number t of unbounded fan-in gates on the input-output path of a Boolean circuit specifying the problem. We have:

$$FPT \subseteq W[1] \subseteq W[2] \subseteq \ldots \subseteq \ldots$$

For instance, the clique problem is $W[1]$-complete with respect to the size of the clique, whilst the dominating set problem is $W[2]$-complete with respect to the size of the dominating set. There is considerable evidence to suggest that $W[1]$-hardness implies parametric intractability. In particular, the halting problem for non-deterministic Turing machines is $W[1]$-complete with respect to the length of the accepting computation.

Returning to the computational complexity of control, both fixed-parameter tractability and intractability results are known. For instance, the control of Llull and Copeland elections by the addition or deletion of voters is fixed-parameter tractable when we bound either the number of candidates or the number of voters [66]. On the other hand, both destructive and constructive control of plurality voting by adding candidates are fixed-parameter intractable (specifically $W[2]$-hard) with respect to the number of added candidates [17]. Such results give additional insight into the source of computational complexity, and they may help understand whether complexity is an issue in a particular case or not.

4.3   MECHANISM DESIGN

Another direction in which to study strategic voting comes from game theory [176]. For example, *mechanism design* considers the design of systems in which agents are motivated to reveal private information like their preferences [131]. A game is *incentive compatible* if all of the participants do best when they truthfully reveal any private information asked for by the mechanism. Any manipulable voting system is not incentive compatible. On the other hand, if votes are single-peaked, then the median voting rule is incentive compatible.

This suggests another escape from manipulation. The Gibbard-Sattertwhaite theorem tells us that truth-telling is not a dominant strategy. That is, it may not be the best strategy for one agent to reveal their true preferences irrespective of how the other agents play. We might instead look for a weaker condition. For example, are there situations where truth-telling is merely a Nash equilibrium? A Nash equilibrium is a situation where there is no incentive for an agent to change their strategy unilaterally [111].

4.4   INCOMPARABILITY

Social choice theory typically assumes preferences are total orders. For instance, Arrow's theorem applies to aggregating totally ordered preferences. However, as we have seen in earlier chapters, preferences and preference formalisms can often represent partial orders in which not all outcomes are ordered but we have incomparability between certain outcomes. As Kelly noticed in [114],

> *"completeness is not an innocuous assumption'...[since a person could be asked] to judge between two alternatives so unlike one another he just can't compare them".*

For example, suppose we are considering different places to rent. While it is easy and reasonable to compare two apartments, it may be difficult to compare an apartment and a house. We may wish simply to declare them incomparable. Moreover, we may have several possibly conflicting preference criteria, and their combination can naturally lead to an incomplete order. For example, we may want a cheap but large apartment, so an 80 square metre apartment which costs 100,000 euros is in some sense incomparable to a 50 square metre apartment which costs 60,000 euros.

Many of the ideas and results discussed above can be generalized from total orders to partial orders. For example, the definitions of strategy proofness, surjectivity, and dictators can be generalized to partial orders. A version of the Gibbard-Sattertwhaite theorem can then be given for partial orders [154]. This is perhaps a little disappointing. Impossibility results like Arrow's and Gibbard-Sattertwhaite's theorem are connected to the problem of having to break Condorcet cycles in some (unfair) way when totally ordering the result. Partial orders would appear to offer us the escape by declaring such outcomes incomparable. However, incomparability on its own is not enough to avoid results like Arrow's or Gibbard-Sattertwhaite's theorems.

4.5   UNCERTAINTY IN PREFERENCE AGGREGATION

4.5.1   INCOMPLETE PROFILES

Another issue with aggregating preferences of multiple voters is eliciting those preferences. As discussed in Chapter 3 for single-agent settings, eliciting preferences takes time and effort. In addition, voters may be reluctant to reveal all their preferences due to privacy and other concerns. We therefore may want to stop elicitation as soon as one candidate has enough support that they must win regardless of any missing preferences. We can consider the computational problem of deciding when we can stop eliciting preferences [41, 187]. This problem has been considered along a number of dimensions: weighted or unweighted votes, bounded or unbounded number of candidates, and *coarse* or *fine* elicitation (i.e., do we elicit a total ordering over the candidates from each voter, or a voter's preferences between two candidates?). For example, deciding if coarse elicitation is over is coNP-complete for STV with unweighted votes and an unbounded number of candidates but polynomial for plurality and Borda [41]. As a second example, for the cup rule on weighted votes, deciding if fine elicitation is over is coNP-complete when there are 4 or more candidates, whilst deciding if coarse elicitation is over is polynomial irrespective of the number of candidates [187]. This demonstrates that what we ask voters can be critical.

With partially elicited preferences, we can define the *possible winners* (those candidates who could still possibly win) as well as the *necessary winner* (the candidate who must now necessarily win) [117]. The necessary winner is always a member of the set of possible winners. When the possible winners equals the unit set containing the necessary winner, we can stop eliciting preferences. We can consider the computational problem of computing these sets [186]. Again, this has been considered along a number of dimensions like weighted or unweighted votes and a bounded or unbounded number of candidates. For example, with weighted votes, computing the set of possible winners for the Borda, Copeland or STV rules is NP-hard [148]. Another setting is when the elicited preferences are just partial orders and we want to compute if a candidate can or must win in all linearizations [192].

The problems of computing necessary and possible winners can be in different complexity classes. For instance, computing the set of possible winners for the second-order Copeland rule with unweighted votes and an unbounded number of candidates is NP-hard, whilst computing the necessary winner is polynomial. In addition, even computing good approximations to the set of possible or necessary winners can be shown to be NP-hard [148]. However, in those cases like the Borda rule where computing the set of possible winners is polynomial, we can focus preference elicitation on those candidates that can still win [148]. This can potentially reduce the number of questions that we must ask the voters.

4.5.2 UNKNOWN AGENDA

Preference elicitation can be useful when we have uncertainty in the votes. There are, however, other types of uncertainty in the context of multi-agent preference aggregation, and the notions of possible and necessary winners are useful also in other situations involving uncertainty. For example, when using the cup rule, there might be uncertainty about the agenda to be used [153]. Indeed, the chair might be keeping the agenda secret to make manipulation more difficult. In addition, we may also have uncertainty about how some of the agents have voted. With both forms of uncertainty, there are several types of possible and necessary winners. Following [153], we say that a candidate is a *weak Condorcet winner* if and only if there is a way for the remaining agents to vote such that the candidate wins whatever the agenda; a candidate is a *strong Condorcet winner* if and only if, however the remaining agents vote, the candidate wins whatever the agenda; a candidate is a *weak possible winner* if and only if there is an agenda and a way for the remaining agents to vote such that the candidate wins; finally, a candidate is a *strong possible winner* if and only if there is an agenda such that however the remaining agents vote the candidate wins. We can also insist that the agenda is balanced (that is, the candidate competing in the most number of pairwise competitions has at most one more competition than the candidate with the least). There are several other dimensions along which we can consider these issues. Are the votes weighted? Are votes represented explicitly or more compactly but perhaps with a loss of information by means of a majority graph (a directed graph in which there is an edge between two candidates if one beats the other in a pairwise comparison)? In [153], it is shown that, if we consider votes given by a majority graph, no matter whether the agenda is known or not, it is polynomial to compute all these notions of winners. With weighted (and possibly incomplete) profiles, weak Condorcet and strong Condorcet winners are also easy to compute. However, determining weak possible winners is NP-complete, even if restrict ourselves to balanced agendas.

4.6 PREFERENCE COMPILATION

Two other computational problems in preference aggregation are how we *communicate* and *compile* the votes. There are two related problems. First, how much information do agents need to communicate between each other in order to compute the winner? Second, supposing the agents declare their preferences in some order, how do we compile the votes cast so far in order to be able to compute the winner? Communication complexity has been used to address the first question. Upper and lower bounds on the communication complexity have been determined for many voting rules [43]. Upper bounds are typically given by the maximum number of bits communicated by a particular deterministic protocol in order for the winner to be computed, whilst the lower bound indicates the minimum requirements of even a non-deterministic protocol. In many cases, the two bounds match indicating that the given deterministic protocol is optimal. For instance, the lower bound for plurality is $\Omega(n \log m)$ bits whilst there is an optimal upper bound of $O(n \log m)$ bits. On the other hand, for STV and Simpson, there are gaps between the lower bound and the best known upper bound. For example, the lower bound for STV is also $\Omega(n \log m)$ bits whilst the best known upper bound is currently $O(n(\log m)^2)$ bits.

Compilation complexity has been used to address the second question. This gives the minimum number of bits required to summarize the votes for a particular rule. Interestingly, voting rules can be ranked quite differently by their compilation and communication complexities. For example, the compilation complexity of plurality with runoff is higher than that of Borda, whilst it has a lower communication complexity [37].

4.7   COMBINATORIAL DOMAINS

When the set of candidates is small, agents may just specify an order over them to express their preferences. However, when the set of candidates is very large, this is infeasible. Large candidate sets occur, for example, when considering the election of committees (rather than single winners), or when voters are presented with multiple referenda.

Actually, situations in which the candidate set is very large are much more frequent than one could expect. For example, if three friends want to decide what to have for dinner and there are four possible options for each of the four components of the menu (first course, main course, dessert and drink), there are 256 possible dinners from which to choose from. As in this case, however, when the set of candidates is very large, it is often possible to see each candidate as the combination of certain features, where each feature has a set of possible instances. This occurs in several AI applications, such as combinatorial auctions, web recommender systems, and configuration systems. Fortunately, in the presence of such a combinatorial structure, agents may describe their preferences in a compact and efficient way, as it has been discussed in Chapter 2.

The decomposition of the candidate set into a combinatorial structure does not come free of cost. Let us consider the following scenario in which there are 13 voters and the candidate set is the Cartesian product of the domain of three binary features, each with domain {*yes*, *no*}. Suppose the votes cast are as follows:

- 3 votes for each of the following: (*yes*, *no*, *no*), (*no*, *yes*, *no*), (*no*, *no*, *yes*);

- 1 vote for each of the following: (*yes*, *yes*, *yes*), (*yes*, *yes*, *no*), (*yes*, *no*, *yes*), (*no*, *yes*, *yes*);

- no vote for (*no*, *no*, *no*).

Suppose we use Plurality to vote on each issue separately in order to obtain the winning combination. Surprisingly, (*no*, *no*, *no*) wins, since 7 out of 13 vote *no* on each issue. This is an instance of the paradox of multiple elections: the winning combination may receive the fewest number of votes [31].

There are several ways to compute a winning candidate when domains are combinatorial [124]. One possibility is to use a voting procedure after having expanded the preferences over the complete set of candidates for all the voters. This option has many drawbacks. It is very costly in terms of space, and, thus, it will only be possible in very small domains, and certainly not when the voting procedure requires a complete ranking of all the candidates (such as the Borda rule). Moreover, the preference orderings induced will most probably have many ties, which in many cases complicates the voting procedure requiring tie-breaking steps.

The space issue can be dealt with by requiring the voters to rank only their k most preferred combinations. However, this may lead to almost random decisions, unless domains are fairly small and there are many voters. For example, if we have 10 binary features, then there are 1024 candidates. Suppose we have 100 voters and we use the Plurality rule, then (relatively) few of these combinations can receive many votes.

Another option is to ask the voters to directly report their compactly represented preferences as ballots and then to apply the voting procedure directly on such succinctly encoded ballots. This is called combinatorial voting [122]. While appealing, this line of research has remained mostly unexplored. It was in fact shown that it is often hard in terms of relevance and computational complexity to apply in a straightforward way voting rules on problems having a combinatorial structure.

Others have investigated conditions under which voting separately on each issue does not lead to paradoxes [195]. Paradoxes, like the one described above, can never be avoided in general. One case where voting on each issue separately is safe is when the preference of voters are preferentially independent. This means that the preferences over one feature are always independent of how the other features are instantiated. Unfortunately, this is a strong assumption.

One way to avoid imposing such a strong restriction is to investigate other ways of aggregating the ballot information coming from feature-by-feature elections. Such an aggregation procedure combines the results obtained on each feature minimizing the distance of the winner from the combinations most preferred by each voter. This approach, however, gives rise to two issues: which metric should be used to measure the distance between the two winners? Also, how the distances to individual ballots should be aggregated? As far as the first issue, one possibility is to use the Hamming distance as a distance metric, thus taking into account the number of features on which the winners differ. As far as the distance aggregation, the most popular approaches are the *Minsum procedure*, where winners that minimize the sum of distances are elected, and the *Minimax procedure*, where winners that minimize the maximal distance to any ballot are elected [30].

Another interesting way to combine elections on single features is to so by voting sequentially. The key idea is to embed the results of previous elections into the voters preferences in order to allow them make their vote for one feature dependent on other features already decided. This approach is promising since it has been shown that it can avoid the multiple-election paradox by preventing candidates that are not preferred by any voter to be elected [121]. It has been investigated both when voters express their preferences as CP-nets [125] as well as with soft constraints [156]. In both cases, several restrictions have been imposed both on the preference structures of each voter in order to ensure properties such as Condorcet consistency and tractability of the winner computation. Despite such restrictions, it has been shown that sequential voting cannot prevent the sequential rule from losing some properties even when they hold for all the rules applied locally to each feature. Neutrality is one such example. However, recent experimental studies conducted with soft constraints suggest that the winner computed sequentially and the one computed in the usual way, by first expanding the voters preferences, do not differ much in terms of the average distance from each voters most preferred candidates [155, 156].

4.8   CONCLUSIONS

Voting is a general mechanism to combine the preferences of multiple agents. Social choice theory, which studies the properties of voting rules, is therefore of relevance to understanding a number of important issues in preference aggregation. However, the computational setting of many preference aggregation problems, and new features like the potentially very large number of candidates gives rise to a number of computational questions. For instance, what is the computational complexity of computing a strategic vote? How do we compute who can still possibly win? Can we design a voting rule where voters are encouraged to be truthful in declaring their preferences? Whilst there has been considerable progress made in addressing these sort of computational questions, research in this area remains very active.

CHAPTER 5

Stable Marriage Problems

Until now, we have considered situations where agents express their preferences over a set of objects. Even in the multi-agent case, such objects are different from the agents. We now consider a slightly different context where the agents express preferences over each other. For example, the agents can be divided in two sets, and each agent in one set expresses its preferences over the agents in the other set. The goal is now to find a "suitable" mapping between the agents of the two sets. From a mathematical point of view, this is called a stable matching problem [168].

Historically, a terminology based on men and women is used to define the problem: given n men and n women, where each person has ranked all members of the opposite sex, we want to marry the men and women together such that there are no two people of opposite sex who would both rather be matched to each other than to their current partners. A pair of a man and woman who are not married to each other who both prefer each other more than their current partner is called a *blocking pair*. If there is no blocking pair, the matching (meaning the set of all man-woman marriages) is said to be *stable*. This problem is usually called the *stable marriage problem* (SMP) [98].

For example, assume we have 3 men and 3 women, and let the set of women be $W = \{w_1, w_2, w_3\}$ and the set of men be $M = \{m_1, m_2, m_3\}$. The following sequence of strict total orders defines the preferences of all men and women (usually called a *profile*):

$m_1 : w_1 \succ w_2 \succ w_3$ (i.e., man m_1 prefers woman w_1 to w_2 to w_3)

$m_2 : w_2 \succ w_1 \succ w_3$

$m_3 : w_3 \succ w_2 \succ w_1$

$w_1 : m_1 \succ m_2 \succ m_3$

$w_2 : m_3 \succ m_1 \succ m_2$

$w_3 : m_2 \succ m_1 \succ m_3$

For this profile, the matching $\{(m_1, w_1), (m_2, w_2), (m_3, w_3)\}$ is stable. In fact, no man and woman, which are not a pair in this matching, form a blocking pair. For example, if we consider m_3 and w_2, we have that w_2 prefers m_3 to her current partner (m_2), but m_3 prefers his current partner to w_2. On the other hand, the matching $\{(m_1, w_2), (m_2, w_1), (m_3, w_3)\}$ is not stable, since the pair (m_1, w_1) is a blocking pair: m_1 prefers w_1 to his current partner (which is w_2), and w_1 prefers m_1 to his current partner (which is m_2).

This abstract problem has a large number of practical applications: assigning medical residents to hospitals [166], students to schools [179], producers to consumers, employers to projects, and so on. Variants of the stable marriage problem turn up in many domains. For example, the US Navy has a web-based multi-agent system for assigning sailors to ships [127]. A list of real-life applications of the stable marriage problem can be found in [197].

5.1   STABILITY

Given a stable marriage problem, Gale and Shapley [82] proved that it is always possible to solve the SMP and find a stable matching. In fact, they provide a quadratic time algorithm which can be used to find one of two particular but extreme stable matchings, the so-called *male optimal* or *female optimal* solutions. S table matchings for an SMP form a lattice with respect to the men's or women's preferences: in this lattice, a stable matching is above another if men in the first matching are matched to equally or more preferred women. Therefore, the top of this lattice is the stable matching where men are mostly satisfied. Conversely, the bottom is the stable matching where men's preferences are least satisfied. For the SMP of the previous section, $\{(m_1, w_1), (m_2, w_2), (m_3, w_3)\}$ is the male-optimal stable marriage, while $\{(w_1, m_1), (w_2, m_3), (w_3, m_2)\}$ is the female-optimal stable marriage.

At the start of the *Gale-Shapley* algorithm [82, 167], each person is free and becomes engaged during the execution of the algorithm. Once a woman is engaged, she never becomes free again (although to whom she is engaged may change), but men can alternate between being free and being engaged. The following step is iterated until all men are engaged: choose a free man m, and have m propose to the most preferred woman w on his preference list, such that w has not already rejected m. If w is free, then w and m become engaged. If w is engaged to man m', then she rejects the man (m or m') that she least prefers, and becomes, or remains, engaged to the other man. The rejected man becomes, or remains, free. When all men are engaged, the engaged pairs form the male optimal stable matching. Figure 5.1 gives the pseudo-code of the Gale-Shapley algorithm.

Gale-Shapley algorithm(a stable marriage problem)
1   Set all men and women as free
2   **while** there is a free man m
3   **do** $w \leftarrow$ the first woman in pref(m) to which he has not yet proposed
4   **if** w is free
5   **then** match m with w
6   **else if** $m \succ_{pref(w)} z$, where z is w's partner
7   **then** match m with w and set z free
8   **else** w rejects m and m remains free

Figure 5.1: The Gale-Shapley algorithm: given an instance of the stable marriage problem, it returns the male-optimal stable marriage. In this pseudo-code, *pref(x)* is the preference list of x.

This algorithm needs a number of steps that is quadratic in n, and it guarantees the following properties:

- If the number of men and women coincide, and all participants express a linear order over all the members of the other group, everyone gets married. Once a woman becomes engaged, she is always engaged to someone. So, at the end, there cannot be a man and a woman both un-engaged, as he must have proposed to her at some point (since a man will eventually propose to every woman, if necessary) and, being un-engaged, she would have to have said yes.

- The returned matching is stable. Let Alice be a woman and Bob be a man. Suppose they are each married, but not to each other. Upon completion of the algorithm, it is not possible for both Alice and Bob to prefer each other over their current partners. If Bob prefers Alice to his current partner, he must have proposed to Alice before he proposed to his current partner. If Alice accepted his proposal, yet is not married to him at the end, she must have rejected him for someone she likes more, and therefore doesn't like Bob more than her current partner. If Alice rejected his proposal, she was already with someone she liked more than Bob.

This algorithm can be improved to avoid redundant proposals that are rejected. Since a woman will never accept a new proposal from a man who is worse than the one she is currently engaged with, it is useless to leave such men in her preference lists. Thus, they are deleted. At the same time, the woman is deleted from these men's preference lists. This leads to the so-called Extended Gale-Shapley (EGS) algorithm [98], where a proposal cannot be rejected, since, if a proposing man m has woman w in his preference list, it means that either w is still single, or she is engaged with a man worse then m (and thus she will accept m's proposal).

As noted above, the Gale-Shapley algorithm always returns a stable marriage. This is, of course, an essential property for any stable matching procedure. However, there are other interesting properties that a stable marriage procedure might have, such as the following:

Gender neutrality: If we swap the role of men and women in the stable marriage procedure, we get the same result.

Peer indifference: The result is not affected by the order in which the members of the same sex are considered.

The Gale-Shapley procedure is peer indifferent, but it is not gender neutral. In fact, if we swap men and women in the example above, we obtain the female optimal solution rather than the male optimal one. Male-optimality can be considered unfair to the women. For this reason, other proposal-based algorithms to compute stable matchings have been proposed that are fairer. An example of such an algorithm is one that computes the minimum-regret stable matching [97]. This is the best stable matching as measured by the person who has the largest regret in it. The regret of a man in a matching is the distance in his preference ordering between his most preferred woman and the woman married to him. The regret of a woman is defined analogously. The algorithm proposed by Gusfield in [97] to compute the minimum-regret stable matching takes $O(n^2)$ time and proceeds by passing from one matching to another. In each step, the person with the maximum regret is identified, and its current marriage is broken in order to obtain another matching with a smaller maximum regret.

5.2   TIES AND INCOMPLETENESS

Often it is not easy to strictly order all items, so we may have ties among men or women in the agents' preference lists [107, 109, 130]. In this case, the classical notion of stability cannot be used and there are alternative notions, among which the most used are the following:

Weak stability: There is no man and no woman, not married to each other, who *strictly* prefer each other to their partners;

Strong stability: There is no man and no woman, not married to each other, such that one strictly prefers the other to its current partner, while the second one likes the first at least as much as its current partner.

Weakly stable matchings always exist. An easy way to find one is to break all ties arbitrarily and then apply the Gale-Shapley algorithm. This will return a weakly stable matching of the original stable marriage problem. On the other hand, strong stable matchings may not exist. An algorithm for deciding existence runs in $O(n^4)$ where n is the number of men.

Another strong assumption in the stable marriage model we have considered so far is that all men rank all women, and vice versa. In many real-life examples, it may be unacceptable for a man (resp., woman) to be paired to a specific woman (resp., man), so this man will not rank her, meaning that he prefers being single to being married to her (resp., him). This means that there could be scenarios where the preference lists are incomplete [98]. In this case, the notion of blocking pair (which makes a marriage unstable) is a man m and a woman w such that:

- m has included w in his preference list;

- w has included m in her preference list;

- m is single or prefers w to his current partner;

- w is single or prefers m to his current partner.

As in the classical case, in stable marriage problems with incomplete lists there is always a stable marriage. Moreover, men and women are partitioned in two sets: those who have partners in all stable marriages, and those who are single. That is, all stable marriages contain the same singles. To find one such stable marriage, it is enough to use the usual Gale-Shapley algorithm, which extends easily to this case.

The scenario becomes more complex when we allow both ties and incompleteness in the preference lists [110]. In this case, weakly stable marriages always exist, but different stable marriages may have different sizes. If one simply wants some stable marriage, again applying the Gale-Shapley algorithm will suffice. However, what one usually wants, in this case, is to find the stable marriage with the minimum number of singles. This problem is NP-hard even if only the women declare ties. It is also NP-hard to approximate a maximum and minimum cardinality stable matching, under certain restrictions [99].

5.3   MANIPULATION

As with voting rules (see Chapter 4), one important issue in procedures to find a stable matching is whether the outcome can be manipulated, that is, whether agents have an incentive to tell the truth or can manipulate the result by mis-reporting their preferences. Unfortunately, Roth [165] has proved that *all* stable marriage procedures can be manipulated, by showing a stable marriage problem with 3 men and 3 women, which can be manipulated whatever stable marriage procedure we use. This result is, in some sense, analogous to the Gibbard-Satterthwaite's theorem [93, 172] for voting theory, already mentioned in Chapter 4, which states that all voting procedures are manipulable under modest assumptions, provided we have 3 or more voters.

A survey of several results about manipulation of stable marriage procedures can be found in [105]. In particular, several early results [51, 56, 84, 165] indicated the futility of a single man lying in the men-proposing algorithm, focusing later work mostly on strategies in which the women lie. However, in [105], it is shown how a coalition of men can get a better result in the men-proposing Gale-Shapley algorithm.

Besides modifying the preference ordering, another way to lie is to truncate the preference lists. Gale and Sotomayor [83] presented a manipulation strategy in which women truncate their preference lists. Roth and Vate [164] discussed strategic issues when the stable matching is chosen at random, proposed a truncation strategy and showed that every stable matching can be achieved as an equilibrium in truncation strategies. Immorlica and Mahdian [106] showed that, if men have preference lists of constant size, while women have complete lists, and both are drawn from an arbitrary distribution of preference lists, the chance of women gaining from lying is vanishingly small. Teo et al. [179] suggested lying strategies for an individual woman, and they proposed an algorithm to find the best partner with the male optimal procedure.

As discussed already in Chapter 4, in voting theory Bartholdi, Tovey and Trick [11] proposed that computational complexity might offer an escape from manipulation: whilst manipulation is always possible, there are voting rules where it is NP-hard to find howw to manipulate. One might hope that computational complexity might also be a barrier to manipulation in stable marriage procedures.

Unfortunately, the men-proposing Gale-Shapley algorithm is computationally easy to manipulate by women [179] (men do not have an incentive to manipulate since they already receive their best partner). When truncated lists are allowed, a woman can just truncate her list after her female-optimal partner, and she will surely get this partner. If instead truncated lists are not allowed, and thus the only form of manipulation is modifying the preference ordering, a manipulating woman is not guaranteed to be able to get her female-optimal partner. However, in polynomial time, she can work out how to reorder her preference list to get her best feasible partner.

If we don't put any restriction on the profiles, there are stable marriage procedures that are NP-hard to manipulate. The intuition is to consider a stable marriage procedure that is computationally easy to compute but NP-hard to invert. To manipulate, a man or a woman will essentially need to be able to invert the procedure to choose between the exponential number of possible preference orderings. Hence, the constructed stable marriage procedure will be NP-hard to manipulate. An example of such a procedure is given in [149].

Another way to construct a stable marriage procedure which is NP-hard to manipulate is to exploit results about the computational complexity of manipulation in voting rules. To define such a stable marriage procedure, we can start by choosing a voting rule (more precisely, a social welfare function) and using it to order the men using the women's preferences and the women using the men's preferences. That is, the social welfare function takes the women's preferences and returns a ranking of the men, and the same for the women. We then construct a male score vector for a marriage using this ordering of the men (where a more preferred man is before a less preferred one): the ith element of the male score vector is the integer j iff the ith man in this order is married to his jth most preferred woman. A large male score vector is a measure of dissatisfaction with the matching from the perspective of the more preferred men. A female score vector is computed in an analogous manner. Then, the overall score for a matching is the lexicographically largest of its male and female score vectors. A large overall score corresponds to dissatisfaction with the matching from the perspective of the more preferred men or women. We then choose the stable matching from our given set which has the lexicographically least overall score. That is, we choose the stable matching which carries less regret for the more preferred men and women. It is NP-complete to decide if an agent can manipulate this marriage procedure when the chosen voting rule is STV [149]. The same holds also when the voting rule is the hybrid plurality rule (one round of the cup rule and then the plurality rule [42]).

5.4   EXTENSIONS

While in the classical formulation, there is the same number of men and women, often in practice, the two sets have a different cardinality. Also, in some cases, we need to match each member of one set with more than one member of the other set, such as in [108], where medical students are matched to pairs of hospitals. In other domains, there could be one set of agents only, each of which expresses its preferences over all others. This is typical, for example, in setting up a team for a project, where each team member expresses its preferences over the other members of the team, as well as in deciding, among a set of students, which pairs of students will share a room. For this last example, this version of the stable marriage problem is called the *stable roommate problem* [38].

5.5   COMPACT PREFERENCE REPRESENTATION

In some applications, the number of men and women can be large. Even if in some domains preference lists may be truncated, it may still be unreasonable to assume that each man and woman provides a strict ordering of many members of the other sex. In addition, eliciting their preferences may be a costly and time-consuming process. This may happen, for example, when men and women have a combinatorial structure, that is, each man and woman may be described by a set of features, where each feature may have a few possible instances, and thus the sets of men and women are the Cartesian product of the sets of their features. In this case, even if the number of features and feature instances is small, the number or men and women can be very large. Also, if there are such features, it may be desirable to express preferences over the feature instances, rather than on specific men or women. For instance, consider a large set of hospitals offering residencies. Doctors looking for such a post might not wish to rank all of the hospitals explicitly, but rather to express their preferences over some features. For example, they might say "I prefer a position close to my home town", or "If the hospital is far away from my home town, then I want a better salary". We should rank the hospitals based on this information.

In this scenario, it may reasonable to use a compact preference formalism, such as soft constraints or CP-nets (see Chapter 2) to express the preference list of each man and woman. The question is whether such a change in preference modelling modifies the computational results or the other properties of the various stable marriage procedures.

In algorithms such as Gale-Shapley's [82] and Gusfield's [97], men make proposals, starting from their most preferred woman and moving down their ordering, whilst women receive proposals and compare these against the men to whom they are currently engaged. Moreover, in both algorithms, proposals are made in increasing order of regret. This is exploited especially in Gusfield's algorithm, where the notion of regret is also used also to decide how to modify the current matching in order to obtain one with a smaller regret. Three operations are thus needed:

- Given a man m, we must find his optimal woman. This is needed the first time a man makes a proposal.

- Given a man m and a woman w, we must find the next best woman for m. This is needed when a man makes a new proposal.

- Given a woman w and two men m_1 and m_2, we must say if m_2 is preferred to m_1 for w. This is needed when a woman compares two proposals to decide whether to remain with the current man (m_1) or to leave him for a new man who is proposing (m_2).

If preferences are given explicitly as strict total orders, as in the traditional SMP setting, these operations take constant time. However, if preferences are represented with a compact representation language, such as CP-nets or soft constraints, then this is not the case in general. In particular, as noted in Chapter 2, finding the optimal solution is easy in an acyclic CP-net (while it is difficult in the general case), but comparing two solutions is difficult even for acyclic binary CP-nets. As for finding the next solution, it is easy in acyclic CP-nets, tree-like CSPs and fuzzy CSPs, but it is difficult in weighted CSPs [28].

It is therefore feasible to consider running the Gale Shapley algorithm and computing the next best woman, or comparing two men, whenever they are needed by the algorithm, rather than computing the full preference orderings upfront. Experimental tests show that this is indeed a reasonable approach, since the Gale-Shapley algorithm needs to consider $O(n^2)$ proposals in the worst case but may, in practice, require only a much smaller number of proposals [147].

5.6   CONSTRAINT-BASED FORMALIZATIONS

The problem of finding a stable matching can be seen as a constraint satisfaction problem: the variables and domains are used to model men and women, while the constraints model the stability condition (besides the obvious matching condition that each man should be paired with a different woman and vice versa). It is therefore natural to investigate the use of the constraint programming machinery to solve stable marriage problems. This may appear overly complex considering that some versions of the stable marriage problem can be solved in polynomial time, but it can instead be very convenient when tackling one of the NP-hard versions.

In [91], it is shown how to model a stable marriage problem (with or without incomplete preference lists) as a constraint satisfaction problem, and it is proven that arc-consistency on the constraint problem has the same effect as the extended Gale-Shapley algorithm on the preference lists. Thus, arc-consistency is enough to obtain the male and the female optimal stable matchings, with no need to use search. It is also shown that search is not needed to enumerate all stable matchings. This shows that generic constraint solving algorithms are equivalent to specialized algorithm for stable marriage problems. In [3], the encoding is done via linear inequalities, while in [183, 184] specialized binary constraints are defined to model stable marriage problems.

NP-hard versions of the stable marriage problem, with both ties and incomplete lists, has been considered in [92], where a complete constraint model is given for such problems. Experimental results show that this encoding, together with off-the-shelf constraint solving technology, is promising in dealing with this general and difficult version of the stable marriage problem.

Other approaches have used non-systematic algorithms based on local search. In [86, 87], there is no need to encode a stable marriage problem as a constraint satisfaction problem since the local search algorithm moves from one matching to the next one by removing a blocking pair (in the case of stable marriage problems, possibly with ties or incomplete lists) or by trying to both minimize blocking pairs and the number of singles (in the case of stable marriage problems with both ties and incomplete lists). While in theory this local search approach is not guaranteed to return a stable matching, or a stable matching with largest size, in practice it is very efficient and almost always finds a matching with the desired properties.

5.7   CONCLUSIONS

Stable marriage problems have a very large number of application domains, from matching students to schools, residents to hospitals, sailors to ships, as well as kidney allocation protocols. In all such applications, it is crucial to assure the users of the stable marriage algorithm that everything is done in a fair way, as well as that there is no (or little) way to manipulate the result. Thus, an axiomatic approach similar to the one classically used for voting theory would be advisable. Related to this, it seems promising to try to achieve a fruitful cross-fertilization between voting theory and stable marriages, since they model different scenarios but have many issues in common. Moreover, compact ways of modelling the agents' preferences should be provided, especially when the number of options (that is, the number of men and women) is large and has a combinatorial structure. In this respect, preference modelling frameworks such as those described in Chapter 2 can be useful. However, the impact of their use on the properties of a stable marriage procedure should be further investigated.

Bibliography

[1] L. El Ghaoui A. Ben-Tal and A. Nemirovski, editors. *Robust Optimization.* Princeton Series in Applied Mathematics. Princeton University Press, 2009. Cited on page(s) 32

[2] G. Adomavicius and A. Tuzhilin. Toward the next generation of recommender systems: a survey of the state-of-the-art and possible extensions. *IEEE Transactions on Knowledge and Data Engineering*, 17, 6, 2005. DOI: 10.1109/TKDE.2005.99 Cited on page(s) 25

[3] B. Aldershof, O. M. Carducci, and D. C. Lorenc. Refined inequalities for stable marriage. *Constraints*, 4:281–292, 1999. DOI: 10.1023/A:1026453915989 Cited on page(s) 68

[4] J. F. Allen. Maintaining knowledge about temporal intervals. *Communications of the ACM*, 26(1):832–843, 1983. DOI: 10.1145/182.358434 Cited on page(s) 21

[5] L. Amgoud, J.-F. Bonnefon, and H. Prade. An argumentation-based approach to multiple criteria decision. In *Proceedings of the 8th European Conference on Symbolic and Quantitative Approaches to Reasoning with Uncertainty (ECSQARU 2005)*, volume 3571 of *LNCS*, pages 269–280. Springer, 2005. DOI: 10.1007/11518655_24 Cited on page(s) 14

[6] J. Amilhastre, H. Fargier, and P. Marquis. Consistency restoration and explanations in dynamic CSPs - application to configuration. *Artificial Intelligence*, 135(1-2):199–234, 2002. DOI: 10.1016/S0004-3702(01)00162-X Cited on page(s) 23

[7] K. Arrow. *Social Choice and Individual Values.* Yale University Press, New Haven, 1970. Cited on page(s) 46

[8] F. Bacchus and A. J. Grove. Graphical models for preference and utility. In *Proceedings of the Eleventh Annual Conference on Uncertainty in Artificial Intelligence (UAI-1995)*, pages 3–10, 1995. Cited on page(s) 27

[9] S. Badaloni, M. Falda, and M. Giacomin. Integrating quantitative and qualitative constraints in fuzzy temporal networks. *AI Communications*, 17(4):183–272, 2004. Cited on page(s) 21

[10] J.J. Bartholdi and J.B. Orlin. Single transferable vote resists strategic voting. *Social Choice and Welfare*, 8(4):341–354, 1991. DOI: 10.1007/BF00183045 Cited on page(s) 48

[11] J.J. Bartholdi, C.A. Tovey, and M.A. Trick. The computational difficulty of manipulating an election. *Social Choice and Welfare*, 6(3):227–241, 1989. DOI: 10.1007/BF00295861 Cited on page(s) 47, 65

[12] J.J. Bartholdi, C.A. Tovey, and M.A. Trick. Voting schemes for which it can be difficult to tell who won the election. *Social Choice and Welfare*, 6(2):157–165, 1989. DOI: 10.1007/BF00303169 Cited on page(s) 44

[13] J.J. Bartholdi, C.A. Tovey, and M.A. Trick. How hard is it to control an election. *Mathematical and Computer Modeling*, 16(8-9):27–40, 1992. DOI: 10.1016/0895-7177(92)90085-Y Cited on page(s) 51

[14] S. Benferhat, D. Dubois, S. Kaci, and H. Prade. Bipolar representation and fusion of preferences on the possibilistic logic framework. In *Proceedings of the Eight International Conference on Principles and Knowledge Representation and Reasoning (KR-02)*, pages 421–448. Morgan Kaufmann, 2002. Cited on page(s) 15

[15] S. Benferhat, D. Dubois, S. Kaci, and H. Prade. Bipolar possibility theory in preference modeling: Representation, fusion and optimal solutions. *Information Fusion*, 7(1):135–150, 2006. DOI: 10.1016/j.inffus.2005.04.001 Cited on page(s) 15

[16] C. Bessiere. Constraint propagation. In *Handbook of constraint programming*, chapter 3, pages 29–84. Elsevier, 2006. Cited on page(s) 8

[17] N. Betzler and J. Uhlmann. Parameterized complexity of candidate control in elections and related digraph problems. *Theoretical Computer Science*, 410(52):5425–5442, 2009. DOI: 10.1016/j.tcs.2009.05.029 Cited on page(s) 55

[18] S. Bistarelli, P. Codognet, and F. Rossi. Abstracting soft constraints: Framework, properties, examples. *Artificial Intelligence*, 139(2):175–211, 2002. DOI: 10.1016/S0004-3702(02)00208-4 Cited on page(s) 4, 23

[19] S. Bistarelli, U. Montanari, and F. Rossi. Semiring-based constraint satisfaction and optimization. *J. ACM*, 44(2):201–236, 1997. DOI: 10.1145/256303.256306 Cited on page(s) 12

[20] S. Bistarelli, U. Montanari, F. Rossi, T. Schiex, G. Verfaillie, and H. Fargier. Semiring-based CSPs and Valued CSPs: Frameworks, properties, and comparison. *Constraints*, 4(3):199–240, 1999. DOI: 10.1023/A:1026441215081 Cited on page(s) 14

[21] S. Bistarelli, M. S. Pini, F. Rossi, and K. B. Venable. Bipolar preference problems: framework, properties and solving techniques. In *Recent Advances in Constraints (CSCLP 2006)*, volume 4651 of *LNCS*, pages 78–92. Springer, 2006. DOI: 10.1007/978-3-540-73817-6_5 Cited on page(s) 3, 15

[22] C. Boutilier, F. Bacchus, and R. Brafman. UCP-networks: A directed graphical representation of conditional utilities. In *Proc. UAI*, 2001. Cited on page(s) 20, 27

[23] C. Boutilier, R. Brafman, H. Hoos, and D. Poole. Reasoning with conditional ceteris paribus preference statements. In *Proc. UAI*, pages 71–80, 1999. Cited on page(s) 16, 18

[24] C. Boutilier, R. I. Brafman, C. Domshlak, H. H. Hoos, and D. Poole. CP-nets: A tool for representing and reasoning with conditional ceteris paribus preference statements. *J. Artif. Intell. Res. (JAIR)*, 21:135–191, 2004. Cited on page(s) 3, 16, 18

[25] C. Boutilier and H. H. Hoos. Bidding languages for combinatorial auctions. In *Proc. IJCAI (International Joint Conference on Artificial Intelligence)*, pages 1211–1217, 2001. Cited on page(s) 26

[26] R. I. Brafman and Y. Dimopoulos. A new look at the semantics and optimization methods of CP-networks. *Computational Intelligence*, 20(2):218–245, 2004. DOI: 10.1111/j.0824-7935.2004.00236.x Cited on page(s) 19

[27] R. I. Brafman and C. Domshlak. Introducing variable importance tradeoffs into cp-nets. In *UAI*, pages 69–76, 2002. Cited on page(s) 18

[28] R. I. Brafman, F. Rossi, D. Salvagnin, K. Brent Venable, and T. Walsh. Finding the next solution in constraint- and preference-based knowledge representation formalisms. In *Proc. KR 2010*, 2010. Cited on page(s) 68

[29] S. J. Brams and P. C. Fishburn. Chapter 4: Voting procedures. In Amartya K. Sen Kenneth J. Arrow and Kotaro Suzumura, editors, *Handbook of Social Choice and Welfare*, volume 1, pages 173 – 236. Elsevier, 2002. Cited on page(s) 42

[30] S.J. Brams, D.M. Kilgour, and M.R. Sanver. A minimax procedure for electing committees. *Public Choice*, 132:401–420, 2007. DOI: 10.1007/s11127-007-9165-x Cited on page(s) 59

[31] S.J. Brams, D.M. Kilgour, and W.S. Zwicker. The paradox of multiple elections. *Social Choice and Welfare*, 15(2):211–236, 1998. DOI: 10.1007/s003550050101 Cited on page(s) 58

[32] R. Cavallo, D. C. Parkes, Adam I. Juda, Adam Kirsch, Alex Kulesza, SÃ©bastien Lahaie, Benjamin Lubin, Loizos Michael, and Jeffrey Shneidman. TBBL: A tree-based bidding language for iterative combinatorial exchanges. In *In Multidisciplinary Workshop on Advances in Preference Handling (at IJCAI)*, 2005. Cited on page(s) 26

[33] M. Ceberio and F. Modave. Interval-based multicriteria decision making. In *AMAI*, 2004. DOI: 10.1109/NAFIPS.2008.4531298 Cited on page(s) 32

[34] L. Chen and P. Pu. Survey of preference elicitation methods. Technical Report IC/200467, Swiss Federal Institute of Technology in Lausanne (EPFL), 2004. Cited on page(s) 4, 25

[35] Y. Chevaleyre, U. Endriss, J. Lang, and N. Maudet. A short introduction to computational social choice. In J. van Leeuwen, G.F. Italiano, W. van der Hoek, C. Meinel, H. Sack, and F. Plasil, editors, *33rd Conference on Current Trends in Theory and Practice of Computer Science (SOFSEM 2007)*, volume 4362 of *Lecture Notes in Computer Science*, pages 51–69. Springer, 2007. Cited on page(s) 41

[36] Y. Chevaleyre, U. Endriss, J. Lang, and N. Maudet. Preference handling in combinatorial domains: From ai to social choice. *AI Magazine*, 29(4):37–46, 2008. Cited on page(s) 41

[37] Y. Chevaleyre, J. Lang, N. Maudet, and G. Ravilly-Abadie. Compiling the votes of a subelectorate. In C. Boutilier, editor, *Proceedings of the 21st International Joint Conference on Artificial Intelligence (IJCAI-2009)*, pages 97–102, 2009. Cited on page(s) 58

[38] K. Chung. On the existence of stable roommate matching. *Games and economic behavior*, 33:206–230, 2000. DOI: 10.1006/game.1999.0779 Cited on page(s) 66

[39] T. Coleman and V. Teague. On the complexity of manipulating elections. In *Proceedings of the 13th The Australasian Theory Symposium (CATS2007)*, pages 25–33, 2007. Cited on page(s) 49, 54

[40] V. Conitzer and T. Sandholm. Complexity of manipulating elections with few candidates. In *Proceedings of the 18th National Conference on AI (AAAI)*, 2002. Cited on page(s) 48

[41] V. Conitzer and T. Sandholm. Vote elicitation: Complexity and strategy-proofness. In *Proceedings of the 18th National Conference on AI (AAAI)*, 2002. Cited on page(s) 56

[42] V. Conitzer and T. Sandholm. Universal voting protocol tweaks to make manipulation hard. In *Proceedings of 18th IJCAI*, pages 781–788, 2003. Cited on page(s) 52, 66

[43] V. Conitzer and T. Sandholm. Communication complexity of common voting rules. In J. Riedl, M.J. Kearns, and M.K. Reiter, editors, *Proceedings 6th ACM Conference on Electronic Commerce (EC-2005)*, pages 78–87. ACM, 2005. Cited on page(s) 57

[44] V. Conitzer, T. Sandholm, and J. Lang. When are elections with few candidates hard to manipulate. *Journal of the Association for Computing Machinery*, 54, 2007. DOI: 10.1145/1236457.1236461 Cited on page(s) 42, 48, 49, 50

[45] P. Cousot and R. Cousot. Abstract interpretation: A unified lattice model for static analysis of programs by construction or approximation of fixpoints. In *Proceedings of the Fourth ACM Symposium on Principles of Programming Languages (POPL 1977)*, pages 238–252, 1977. DOI: 10.1145/512950.512973 Cited on page(s) 4, 23

[46] S. de Givry, J. Larrosa, P. Meseguer, and T. Schiex. Solving Max-SAT as weighted CSP. In *Proc. CP 2003*, LNCS. Springer Verlag LNCS 2833, 2003. Cited on page(s) 25

[47] R. Dechter. Bucket elimination: A unifying framework for reasoning. *Artificial Intelligence*, 113:41–85, 1999. DOI: 10.1016/S0004-3702(99)00059-4 Cited on page(s) 14

[48] R. Dechter. *Constraint Processing*. Morgan Kaufmann, 2003. Cited on page(s) 4

[49] R. Dechter. Tractable structures for constraint satisfaction problems. In F. Rossi, T. Walsh, and P. van Beek, editors, *Handbook of Constraint Programming*. Elsevier, 2006. Cited on page(s) 5

[50] R. Dechter, I. Meiri, and J. Pearl. Temporal constraint networks. *Artificial Intelligence*, 49:61–95, 1991. DOI: 10.1016/0004-3702(91)90006-6 Cited on page(s) 21

[51] G. Demange, D. Gale, and M. Sotomayor. A further note on the stable matching problem. *Discrete Applied Mathematics*, 16:217–222, 1987. DOI: 10.1016/0166-218X(87)90059-X Cited on page(s) 65

[52] C. Domshlak and R. Brafman. CP-nets — reasoning and consistency testing. In *Proc. KR*, pages 121–132, 2002. Cited on page(s) 19

[53] C. Domshlak, F. Rossi, K. B. Venable, and T. Walsh. Reasoning about soft constraints and conditional preferences: complexity results and approximation techniques. In *Proceedings of the Eighteenth International Joint Conference on Artificial Intelligence (IJCAI 2003)*, pages 215–220. Morgan Kaufmann, 2003. Cited on page(s) 20

[54] R. G. Downey, M. R. Fellows, and U. Stege. Parameterized complexity: A framework for systematically confronting computational intractability. In R. Graham, J. Kratochvil, J. Nesetril, and F. Roberts, editors, *Contemporary Trends in Discrete Mathematics: From DIMACS and DIMATIA to the Future*, volume 49 of *DIMACS Series in Discrete Mathematics and Theoretical Computer Science*, pages 49–99. American Mathematical Society, 1999. Cited on page(s) 54

[55] J. Doyle and M. Wellman. Representing preferences as ceteris paribus comparatives. In *Proc. AAAI Spring Symposium on Decision-Making Planning*, pages 69–75, 1994. Cited on page(s) 16

[56] L. Dubins and D. Freedman. Machiavelli and the Gale-Shapley algorithm. *American Mathematical Monthly*, 88:485–494, 1981. DOI: 10.2307/2321753 Cited on page(s) 65

[57] D. Dubois and H. Fargier. On the qualitative comparison of sets of positive and negative affects. In *Proceedings of the 8th European Conference on Symbolic and Quantitative Approaches to Reasoning with Uncertainty (ECSQARU 2005)*, volume 3571 of *LNCS*, pages 305–316. Springer, 2005. DOI: 10.1007/11518655_27 Cited on page(s) 14

[58] D. Dubois and H. Fargier. Qualitative decision making with bipolar information. In *Proceedings of the Tenth International Conference on Principles of Knowledge Representation and Reasoning (KR-06)*, pages 175–186. AAAI Press, 2006. Cited on page(s) 14

[59] D. Dubois, H. Fargier, and H. Prade. The calculus of fuzzy restrictions as a basis for flexible constraint satisfaction. In *2nd IEEE Int. Conf. on Fuzzy Systems*. IEEE, 1993. DOI: 10.1109/FUZZY.1993.327356 Cited on page(s) 10

[60] C. Dwork, R. Kumar, M. Naor, and D. Sivakumar. Rank aggregation methods for the web. In *Proceedings of the 10th International World-Wide Web Conference (WWW)*, pages 613–622, 2001. DOI: 10.1145/371920.372165 Cited on page(s) 41

[61] M. Ehrgott and X. Gandibleux, editors. *Multiple Criteria Optimization: State of the art annotated bibliographic surveys*. Kluwer Academic, Dordrecht, 2002. Cited on page(s) 15

[62] E. Elkind and H. Lipmaa. Hybrid voting protocols and hardness of manipulation. In *Proceedings of the 16th Annual International Symposium on Algorithms and Computation (ISAAC'05)*, 2005. DOI: 10.1007/11602613_22 Cited on page(s) 53

[63] P. Faliszewski, E. Hemaspaandra, and L.A. Hemaspaandra. How hard is bribery in elections? *J. Artif. Intell. Res. (JAIR)*, 35:485–532, 2009. DOI: 10.1613/jair.2676 Cited on page(s) 52

[64] P. Faliszewski, E. Hemaspaandra, and L.A. Hemaspaandra. Multimode control attacks on elections. In C. Boutilier, editor, *Proceedings of the 21st International Joint Conference on Artificial Intelligence (IJCAI-2009)*, pages 128–133, 2009. Cited on page(s) 52

[65] P. Faliszewski, E. Hemaspaandra, L.A. Hemaspaandra, and J. Rothe. Llull and Copeland voting broadly resist bribery and control. In *Proceedings of the 22nd National Conference on AI*, pages 724–730. Association for Advancement of Artificial Intelligence, 2007. Cited on page(s)

[66] P. Faliszewski, E. Hemaspaandra, L.A. Hemaspaandra, and J. Rothe. Llull and Copeland voting computationally resist bribery and constructive control. *Journal of Artificial Intelligence Research (JAIR)*, 35:275–341, 2009. DOI: 10.1613/jair.2697 Cited on page(s) 43, 55

[67] P. Faliszewski, E. Hemaspaandra, and H. Schnoor. Copeland voting: ties matter. In L. Padgham, D.C. Parkes, J. Müller, and S. Parsons, editors, *7th International Joint Conference on Autonomous Agents and Multiagent Systems (AAMAS 2008)*, pages 983–990, 2008. Cited on page(s) 51

[68] P. Faliszewski and A.D. Procaccia. AI's war on manipulation: Are we winning? *AI Magazine*, 31(4):53–64, 2010. Cited on page(s) 41

[69] H. Fargier and J. Lang. Uncertainty in constraint satisfaction problems: a probabilistic approach. In *Proceedings of the European Conference on Symbolic and Quantitative Approaches to Reasoning and Uncertainty (ECSQARU 1993)*, volume 747 of *LNCS*, pages 97–104. Springer, 1993. DOI: 10.1007/BFb0028188 Cited on page(s) 11

[70] H. Fargier, J. Lang, and T. Schiex. Selecting preferred solutions in fuzzy constraint satisfaction problems. In *Proceedings of the First European Congress on Fuzzy and Intelligent Technologies (EUFIT-93)*. Verlag der Augustinus Buchhandlung, Aachen, 1993. Cited on page(s) 10

[71] P. Fishburn and P. Wakker. The invention of the independence condition for preferences. *Management Science*, 41(7):1130–1144, 1995. DOI: 10.1287/mnsc.41.7.1130 Cited on page(s) 16

[72] P. C. Fishburn. *Utility theory for decision making*. Wiley, 1970. Cited on page(s) 27

[73] P.C. Fishburn. Additive utilities with incomplete product set: Applications to priorities and assignments. *Operations Research*, 15(3):537–542, 1967. DOI: 10.1287/opre.15.3.537 Cited on page(s) 27

[74] E. C. Freuder. A sufficient condition for backtrack-free search. *J. ACM*, 29:24–32, 1982. DOI: 10.1145/322290.322292 Cited on page(s) 5

[75] E. C. Freuder, C. Likitvivatanavong, M. Moretti, F. Rossi, and R. J. Wallace. Computing explanations and implications in preference-based configurators. In *Recent Advances in Constraints (CSCLP 2003)*, volume 2627 of *LNCS*, pages 76–92. Springer, 2003. DOI: 10.1007/3-540-36607-5_6 Cited on page(s) 4, 24

[76] E. C. Freuder, C. Likitvivatanavong, and R. J. Wallace. Deriving explanations and implications for constraint satisfaction problems. In *Proceedings of the 7th International Conference of Principles and Practice of Constraint Programming (CP 2001)*, volume 2239 of *LNCS*, pages 585–589. Springer, 2001. DOI: 10.1007/3-540-45578-7_44 Cited on page(s) 24

[77] E. C. Freuder and D. Sabin. Interchangeability supports abstraction and reformulation for multi-dimensional constraint satisfaction. In *Proceedings of the Fourteenth National Conference on Artificial Intelligence (AAAI-97)*, pages 191–196. AAAI Press / The MIT Press, 1997. Cited on page(s) 23

[78] E. C. Freuder and R. J. Wallace. Partial constraint satisfaction. *Artificial Intelligence*, 58(1-3):21–70, 1992. DOI: 10.1016/0004-3702(92)90004-H Cited on page(s) 12

[79] E. Friedgut, G. Kalai, and N. Nisan. Elections can be manipulated often. In *Proc. 49th FOCS*. IEEE Computer Society Press, 2008. DOI: 10.1109/FOCS.2008.87 Cited on page(s) 53

[80] J. Fürnkranz and E. Hüllermeier, editors. *Preference Learning*. Springer, 2010. Cited on page(s) 25

[81] T. Gal, T.J. Stewart, and T. Hanne, editors. *Multicriteria Decision Making: Advances in MCDM Models, Algorithms, Theory and Applications*, volume 21 of *International Series in Operations Research and Management Science*. Springer, 1999. Cited on page(s) 27

[82] D. Gale and L. S. Shapley. College admissions and the stability of marriage. *American Mathematical Monthly*, 69, 1962. DOI: 10.2307/2312726 Cited on page(s) 62, 67

[83] D. Gale and M. Sotomayor. Machiavelli and the stable matching problem. *American Mathematical Monthly*, 92:261–268, 1985. DOI: 10.2307/2323645 Cited on page(s) 65

[84] D. Gale and M. Sotomayor. Some remarks on the stable matching problem. *Discrete Applied Mathematics*, 11:223–232, 1985. DOI: 10.1016/0166-218X(85)90074-5 Cited on page(s) 65

[85] M. Gelain, M. S. Pini, F. Rossi, and K. B. Venable. Dealing with incomplete preferences in soft constraint problems. In *Proceedings of the 13th International Conference on Principles and Practice of Constraint Programming (CP 2007)*, volume 4741 of *LNCS*, pages 286–300. Springer, 2007. DOI: 10.1007/978-3-540-74970-7_22 Cited on page(s) 4

[86] M. Gelain, M. S. Pini, F. Rossi, K. B. Venable, and T. Walsh. Local search algorithms on the stable marriage problem: Experimental studies. In *Proc. ECAI 2010*, pages 1085–1086. IOS Press, 2010. DOI: 10.3233/978-1-60750-606-5-1085 Cited on page(s) 68

[87] M. Gelain, M. S. Pini, F. Rossi, K. B. Venable, and T. Walsh. Local search for stable marriage problems with ties and incomplete lists. In *Proc. PRICAI 2010*, volume 6230 of *Lecture Notes in Computer Science*, pages 64–75. Springer, 2010. DOI: 10.1007/978-3-642-15246-7_9 Cited on page(s) 68

[88] M. Gelain, M.S. Pini, F. Rossi, K. B. Venable, and T. Walsh. Elicitation strategies for soft constraint problems with missing preferences: Properties, algorithms and experimental studies. *Artificial Intelligence*, 174(3-4):270–294, 2010. DOI: 10.1016/j.artint.2009.11.015 Cited on page(s) 33, 35

[89] M. Gelain, M.S. Pini, F. Rossi, K. B. Venable, and T. Walsh. A local search approach to solve incomplete fuzzy and weighted csps. In *Proceedings the 3rd International Conference on Agents and Artificial Intelligence (ICAART 2011)*. SciTePress, 2011. Cited on page(s) 37

[90] M. Gelain, M.S. Pini, F. Rossi, K.B Venable, and N. Wilson. Interval-valued soft constraint problems. *Annals of Mathematics and Artificial Intelligence (AMAI)*, 58(3-4):261–298, 2010. DOI: 10.1007/s10472-010-9203-0 Cited on page(s) 29

[91] I. P. Gent, R. W. Irving, D. F. Manlove, P. Prosser, and B. M. Smith. A constraint programming approach to the stable marriage problem. In *Proc. CPâŁ™01*, pages 225–239. Springer LNCS 2239, 2001. DOI: 10.1007/3-540-45578-7_16 Cited on page(s) 68

[92] I. P. Gent and P. Prosser. An empirical study of the stable marriage problem with ties and incomplete lists. In *Proc. ECAI 2002*, pages 141–145. IOS Press, 2002. Cited on page(s) 68

[93] A. Gibbard. Manipulation of voting schemes: A general result. *Econometrica*, 41:587–601, 1973. DOI: 10.2307/1914083 Cited on page(s) 47, 65

[94] J. Goldsmith, J. Lang, M. Truszczynski, and N. Wilson. The computational complexity of dominance and consistency in CP-nets. *JAIR*, 33(1):403–432, 2008. DOI: 10.1613/jair.2627 Cited on page(s) 18

[95] M. Grabisch, B. de Baets, and J. Fodor. The quest for rings on bipolar scales. *Int. Journ. of Uncertainty, Fuzziness and Knowledge-Based Systems*, 2003. DOI: 10.1142/S0218488504002941 Cited on page(s) 15

[96] B. Grofman. Black's single-peakedness condition. In C.K. Rowley and F. Schneider, editors, *The Encyclopedia of Public Choice*, pages 367–369. Springer, 2003. Cited on page(s) 47

[97] D. Gusfield. Three fast algorithms for four problems in stable marriage. *SIAM Journal of Computing*, 16(1), 1987. DOI: 10.1137/0216010 Cited on page(s) 63, 64, 67

[98] D. Gusfield and R. W. Irving. *The Stable Marriage Problem: Structure and Algorithms*. MIT Press, 1989. Cited on page(s) 61, 63, 64

[99] M. Halldorsson, R. W. Irving, K. Iwama, D. Manlove, S. Miyazaki, Y. Morita, and S. Scott. Approximability results for stable marriage problems with ties. *Theoretical Computer Science*, 306(1-3):431–447, 2003. DOI: 10.1016/S0304-3975(03)00321-9 Cited on page(s) 65

[100] S. O. Hansson. Preference logic. In *Handbook of Philosophical Logic, vol.4*, pages 319–394. Kluwer, 2001. Cited on page(s) 16

[101] E. Hemaspaandra, L.A. Hemaspaandra, and J. Rothe. Hybrid elections broaden complexity-theoretic resistance to control. *Mathematical Logic Quarterly*, 55(4):397–424, 2009. DOI: 10.1002/malq.200810019 Cited on page(s) 53

[102] F. Heras and J. Larrosa. Resolution in Max-SAT and its relation to local consistency in weighted CSPs. In *Proc. IJCAI 2005*, 2005. Cited on page(s) 25

[103] J. Herlocker, J. Konstan, A. Borchers, and J. Riedl J. An algorithmic framework for performing collaborative filtering. In *Proceedings of the 1999 Conference on Research and Development in Information Retrieval*, 1999. DOI: 10.1145/312624.312682 Cited on page(s) 25

[104] H. H. Hoos and T. Stützle. *Stochastic Local Search: Foundations & Applications*. Elsevier / Morgan Kaufmann, 2004. Cited on page(s) 7

[105] C.-C. Huang. Cheating by men in the Gale-Shapley stable matching algorithm. In *Proc. ESA'06*, pages 418–431. Springer-Verlag, 2006. DOI: 10.1007/11841036_39 Cited on page(s) 65

[106] N. Immorlica and M. Mahdian. Marriage, honesty, and stability. In *Proc. SODA'05*, pages 53–62, 2005. DOI: 10.1145/1070432.1070441 Cited on page(s) 65

[107] R. W. Irving. Stable marriage and indifference. *Discrete Applied Mathematics*, 48:261–272, 1994. DOI: 10.1016/0166-218X(92)00179-P Cited on page(s) 64

[108] R. W. Irving. Matching medical students to pairs of hospitals: a new variation on an old theme. In *Proc. ESA'98*, volume 1461 of *LNCS*, pages 381–392. Springer-Verlag, 1998. DOI: 10.1007/3-540-68530-8_32 Cited on page(s) 66

[109] R. W. Irving, D. Manlove, and S. Scott. The hospital/residents problem with ties. In *Proc. SWATT'00*, volume 1851, pages 259–271. Springer-Verlag, 2000. Cited on page(s) 64

[110] K. Iwama, D. Manlove, S. Miyazaki, and Y. Morita. Stable marriage with incomplete lists and ties. In *Proc. ICALP'99*, volume 1644 of *LNCS*, pages 443–452. Springer, 1999. DOI: 10.1007/3-540-48523-6_41 Cited on page(s) 64

[111] J.F. Nash Jr. Equilibrium points in n-person games. *Proceedings of the National Academy of Sciences of the United States of America*, 36(1):48–49, 1950. DOI: 10.1073/pnas.36.1.48 Cited on page(s) 55

[112] U. Junker. QUICKXPLAIN: Preferred explanations and relaxations for over-constrained problems. In *Proceedings of the Nineteenth National Conference on Artificial Intelligence (AAAI 2004)*, pages 167–172. AAAI Press / The MIT Press, 2004. Cited on page(s) 23

[113] N. Jussien and V. Barichard. The PaLM system: explanation-based constraint programming. In *Proceedings of TRICS: Techniques foR Implementing Constraint programming Systems, a post-conference workshop of CP 2000*, pages 118–133, Singapore, 2000. Cited on page(s) 23

[114] J.S. Kelly. *Social choice theory. An introduction*. Springer-Verlag, 1988. Cited on page(s) 55

[115] L. Khatib, P. H. Morris, R. Morris, F. Rossi, A. Sperduti, and K. B. Venable. Solving and learning a tractable class of soft temporal constraints: Theoretical and experimental results. *AI Commun.*, 20(3):181–209, 2007. Cited on page(s) 21

[116] L. Khatib, P. H. Morris, R. A. Morris, and F. Rossi. Temporal constraint reasoning with preferences. In *Proceedings of the Seventeenth International Joint Conference on Artificial Intelligence (IJCAI 2001)*, pages 322–327. Morgan Kaufmann, 2001. Cited on page(s) 4, 21

[117] K. Konczak and J. Lang. Voting procedures with incomplete preferences. In *Proceedings of the IJCAI-2005 workshop on Advances in Preference Handling*, 2005. Cited on page(s) 56

[118] T. Koski and J. M. Noble. *Bayesian networks: an introduction*. John Wiley and Sons, 2009. Cited on page(s) 16

[119] D. Krantz, R. D. Luce, P. Suppes, and A. Tversky. *Foundations of Measurement Volume I (Additive and polynomial representations)*. Academic Press, 1971. Cited on page(s) 27

[120] C. Labreuche and M. Grabisch. Generalized Choquet-like aggregation functions for handling bipolar scales. *European Journal of Operational Research*, 172(3):931–955, 2006. DOI: 10.1016/j.ejor.2004.11.008 Cited on page(s) 14

[121] D. Lacy and E.M.S. Niou. A problem with referendums. *Journal of Theoretical Politics*, 12(1):5–31, 2000. DOI: 10.1177/0951692800012001001 Cited on page(s) 59

[122] J. Lang. Logical preference representation and combinatorial vote. *Ann. Math. Artif. Intell.*, 42(1-3):37–71, 2004. DOI: 10.1023/B:AMAI.0000034522.25580.09 Cited on page(s) 26, 59

[123] J. Lang. Logical preference representation and combinatorial vote. *Annals of Mathematics and Artificial Intelligence*, 42(1-3):37–71, 2004. DOI: 10.1023/B:AMAI.0000034522.25580.09 Cited on page(s) 41

[124] J. Lang. Voting in combinatorial domains: What logic and AI have to say. In S. Hölldobler, C. Lutz, and H. Wansing, editors, *Logics in Artificial Intelligence, 11th European Conference, (JELIA 2008)*, volume 5293 of *Lecture Notes in Computer Science*, pages 5–7. Springer, 2008. Cited on page(s) 59

[125] J. Lang and L. Xia. Sequential composition of voting rules in multi-issue domains. *Mathematical Social Sciences*, 57(3):304–324, 2009. DOI: 10.1016/j.mathsocsci.2008.12.010 Cited on page(s) 59

[126] C. Lecoutre, S. Merchez, F. Boussemart, and E. Grégoire. A CSP abstraction framework. In *Proceedings of the 4th International Symposium on Abstraction, Reformulation, and Approximation (SARA 2000)*, volume 1864 of *LNCS*. Springer, 2000. DOI: 10.1007/3-540-44914-0_27 Cited on page(s) 23

[127] J. Liebowitz and J. Simien. Computational efficiencies for multi-agents: a look at a multi-agent system for sailor assignment. *Electonic government: an International Journal*, 2(4):384–402, 2005. DOI: 10.1504/EG.2005.008330 Cited on page(s) 62

[128] P.R. Lowrie. The Sydney coordinated adaptive traffic system: Principles, methodology, algorithms. In *Proceedings of International Conference on Road Traffic Signaling*, pages 67–70. Institution of Electrical Engineers, London, U.K., 1982. Cited on page(s) 41

[129] D. Mailharro. A classification and constraint-based framework for configuration. *Artificial Intelligence for Engineering Design, Analysis and Manufacturing*, 12(4):383–397, 1998. Cited on page(s) 24

[130] D. Manlove. The structure of stable marriage with indifference. *Discrete Applied Mathematics*, 122(1-3):167–181, 2002. DOI: 10.1016/S0166-218X(01)00322-5 Cited on page(s) 64

[131] A. Mas-Colell, M.D. Whinston, and J.R. Green. *Microeconomic Theory*. Oxford University Press, New York, 1995. Cited on page(s) 55

[132] K. May. A set of independent necessary and sufficient conditions for simple majority decisions. *Econometrica*, 20(4):680–684, 1952. DOI: 10.2307/1907651 Cited on page(s) 42

[133] P. Melville and V. Sindhwani. Recommender systems. In Claude Sammut and Geoffrey Webb, editors, *Encyclopedia of Machine Learning*. Springer, 2010. Cited on page(s) 25

[134] P. Meseguer, F. Rossi, and T. Schiex. Soft constraints. In *Handbook of constraint programming*, chapter 9, pages 281–328. Elsevier, 2006. DOI: 10.1016/S1574-6526(06)80013-1 Cited on page(s) 3, 12, 14, 22

[135] M. Michalowski, C. A. Knoblock, K. M. Bayer, and B. Y. Choueiry. Exploiting automatically inferred constraint-models for building identification in satellite imagery. In *Proceedings of the 15th ACM International Symposium on Geographic Information Systems,(ACM-GIS 2007)*, page 6. ACM, 2007. DOI: 10.1145/1341012.1341021 Cited on page(s) 25

[136] F. Modave and M. Grabisch. Preferential independence and the Choquet integral. In *Proc. FUR'97*, 1997. Cited on page(s) 27

[137] F. Modave and M. Grabisch. Preference representation by the Choquet integral: the commensurability hypothesis. In *Proc. IPMU 1998*, 1998. Cited on page(s) 27

[138] M. D. Moffitt and M. E. Pollack. Temporal preference optimization as weighted constraint satisfaction. In *Proceedings of the Twenty-First National Conference on Artificial Intelligence (AAAI 2006)*. AAAI Press, 2006. Cited on page(s) 22

[139] H. Moulin. *Axioms of cooperative decision making*. Cambridge University Press, United Kingdom, 1988. Cited on page(s) 45

[140] N. Nisan. Bidding languages for combinatorial auctions. In *Combinatorial auctions*. MIT Press, 2006. Cited on page(s) 26

[141] S. O'Connell, B. O'Sullivan, and E. Freuder. Strategies for interactive constraint acquisition. In *Proceedings of the 8th International Conference on Principles and Practice of Constraint Programming, (CP 2002)*, volume 2470 of *LNCS*. Springer, 2002. Cited on page(s) 25

[142] M. Öztürk and A. Tsoukiàs. Preference representation with 3-points intervals. In *ECAI*, volume 141 of *Frontiers in Artificial Intelligence and Applications*, pages 417–421. IOS Press, 2006. Cited on page(s) 32

[143] J. Pearl. Bayesian inference methods. In *Encyclopedia of Artificial Intelligence*, pages 89–98. John Wiley & Sons, 1992. Cited on page(s) 26

[144] B. Peintner and M. E. Pollack. Low-cost addition of preferences to DTPs and TCSPs. In *Proceedings of the Nineteenth National Conference on Artificial Intelligence, (AAAI 2004)*, pages 723–728. AAAI Press / The MIT Press, 2004. Cited on page(s) 4, 21

[145] B. Peintner and M. E. Pollack. Anytime, complete algorithm for finding utilitarian optimal solutions to STPPs. In *Proceedings of The Twentieth National Conference on Artificial Intelligence (AAAI 2005)*, pages 443–448. AAAI Press / The MIT Press, 2005. Cited on page(s) 22

[146] D.M. Pennock, E. Horvitz, and C. Lee Giles. Social choice theory and recommender systems: Analysis of the axiomatic foundations of collaborative filtering. In *Proceedings of the 17th National Conference on Artificial Intelligence (AAAI-2000)*, pages 729–734. AAAI Press, 2000. Cited on page(s) 41

[147] E. Pilotto, F. Rossi, K. B. Venable, and T. Walsh. Compact preference representation in stable marriage problems. In *Proc. ADT 2009 (1st International Conference on Algorithmic Decision Theory), Venice, Italy, October 2009*. Springer LNAI 5783, 2009.
DOI: 10.1007/978-3-642-04428-1_34 Cited on page(s) 68

[148] M. Pini, F. Rossi, B. Venable, and T. Walsh. Incompleteness and incomparability in preference aggregation. In *Proceedings of 20th IJCAI*, 2007. DOI: 10.1016/j.artint.2010.11.009 Cited on page(s) 56, 57

[149] M. S. Pini, F. Rossi, K. B. Venable, and T. Walsh. Manipulation complexity and gender neutrality in stable marriage procedures. *Journal of Autonomous Agents and Multi-Agent Systems*, 22-1, 2011. DOI: 10.1007/s10458-010-9121-x Cited on page(s) 66

[150] M.S. Pini, F.Rossi, and K.B. Venable. Soft constraint problems with uncontrollable variables. *Journal of Experimental and Theoretical Artificial Intelligence (JETAI)*, 22(4):269–310, 2010. DOI: 10.1080/09528131003712962 Cited on page(s) 39

[151] M.S. Pini, F. Rossi, K.B. Venable, and R. Dechter. Robust solutions in unstable optimization problems. In *CSCLP*, volume 5655 of *Lecture Notes in Computer Science*, pages 116–131. Springer, 2008. DOI: 10.1007/978-3-642-03251-6_8 Cited on page(s) 32

[152] M.S. Pini, F. Rossi, K.B. Venable, and T. Walsh. Aggregating partially ordered preferences. *Journal of Logic and Computation*, 2008. DOI: 10.1093/logcom/exn012 Cited on page(s)

[153] M.S. Pini, F. Rossi, K.B. Venable, and T. Walsh. Dealing with incomplete agents' preferences and an uncertain agenda in group decision making via sequential majority voting. In *11th International Conference on Principles of Knowledge Representation and Reasoning (KR-2008)*. AAAI Press, 2008. A longer version will appear in the Journal of Autonomous Agents and Multi-Agent Systems. Cited on page(s) 52, 57

[154] M.S. Pini, F. Rossi, K.B. Venable, and T. Walsh. Aggregating partially ordered preferences. *Journal of Logic and Computation*, 19(3):475–502, 2009. DOI: 10.1093/logcom/exn012 Cited on page(s) 56

[155] G. Dalla Pozza, M. S. Pini, F. Rossi, and K. B. Venable. Multi-agent soft constraint aggregation via sequential voting. In *Proc. IJCAI 2011*, 2011. Cited on page(s) 60

[156] G. Dalla Pozza, F. Rossi, and K. B. Venable. Multi-agent soft constraint aggregation: a sequential approach. In *Proceedings the 3rd International Conference on Agents and Artificial Intelligence (ICAART 2011)*, volume 1. SciTePress, 2011. Cited on page(s) 59, 60

[157] S. D. Prestwich, F. Rossi, K. B. Venable, and T. Walsh. Constraint-based preferential optimization. In *AAAI*, pages 461–466. AAAI Press / The MIT Press, 2005. Cited on page(s) 20

[158] A. D. Procaccia and J. S. Rosenschein. Average-case tractability of manipulation in voting via the fraction of manipulators. In *Proceedings of 6th Intl. Joint Conference on Autonomous Agents and Multiagent Systems (AAMAS-07)*, pages 718–720, 2007. DOI: 10.1145/1329125.1329255 Cited on page(s) 53

[159] A. D. Procaccia and J. S. Rosenschein. Junta distributions and the average-case complexity of manipulating elections. *Journal of Artificial Intelligence Research*, 28:157–181, 2007. DOI: 10.1145/1160633.1160726 Cited on page(s) 53

[160] P.J. Reny. Arrow's theorem and the Gibbard-Satterthwaite theorem: a unified approach. *Economics Letters*, 70(1):99–105, 2001. DOI: 10.1016/S0165-1765(00)00332-3 Cited on page(s) 46, 47

[161] F. Rossi and A. Sperduti. Learning solution preferences in constraint problems. *Journal of Experimental and Theoreetical Artificial Intelligence*, 10(1):103–116, 1998. DOI: 10.1080/095281398146941 Cited on page(s) 4, 24

[162] F. Rossi and A. Sperduti. Acquiring both constraint and solution preferences in interactive constraint systems. *Constraints*, 9(4):311–332, 2004. DOI: 10.1023/B:CONS.0000049206.43218.5f Cited on page(s) 25

[163] F. Rossi, P. Van Beek, and T. Walsh, editors. *Handbook of Constraint Programming*. Elsevier, 2006. Cited on page(s) 4, 24

[164] A. Roth and V. Vate. Incentives in two-sided matching with random stable mechanisms. *Economic Theory*, 1:31–44, 1991. DOI: 10.1007/BF01210572 Cited on page(s) 65

[165] A. E. Roth. The economics of matching: Stability and incentives. *Mathematics of Operations Research*, 7:617–628, 1982. DOI: 10.1287/moor.7.4.617 Cited on page(s) 65

[166] A. E. Roth. The evolution of the labor market for medical interns and residents: a case study in game theory. *Journal of Political Economy*, 92:991–1016, 1984. DOI: 10.1086/261272 Cited on page(s) 61

[167] A. E. Roth. Deferred acceptance algorithms: History, theory, practice, and open questions. *International Journal of Game Theory, Special Issue in Honor of David Gale on his 85th birthday*, 36:537–569, 2008. DOI: 10.1007/s00182-008-0117-6 Cited on page(s) 62

[168] A. E. Roth and M. Sotomayor. *Two-Sided Matching: A Study in Game-Theoretic Modeling and Analysis*. Cambridge University Press, 1990. Cited on page(s) 61

[169] T. Russell and T. Walsh. Manipulating tournaments in cup and round robin competitions. In Francesca Rossi and Alexis Tsoukiàs, editors, *Algorithmic Decision Theory, First International Conference, ADT 2009, Venice, Italy, October 20-23, 2009. Proceedings*, volume 5783 of *Lecture Notes in Computer Science*, pages 26–37. Springer, 2009. Cited on page(s) 52

[170] Z. Ruttkay. Fuzzy constraint satisfaction. In *3rd IEEE Int. Conf. on Fuzzy Systems*. IEEE, 1994. DOI: 10.1109/FUZZY.1994.343640 Cited on page(s) 10

[171] D. Sabin and R. Weigel. Product configuration frameworks — a survey. *IEEE Intelligent Systems and their Applications*, 13(4):42–49, 1998. DOI: 10.1109/5254.708432 Cited on page(s) 3, 23

[172] M. Satterthwaite. Strategy-proofness and Arrow's conditions: Existence and correspondence theorems for voting procedures and social welfare functions. *Journal of Economic Theory*, 10:187–216, 1975. DOI: 10.1016/0022-0531(75)90050-2 Cited on page(s) 47, 65

[173] J. B. Schafer, J. Konstan, and J. Riedl. Recommender systems in e-commerce. In *Proceedings of the ACM Conference on Electronic Commerce*, 1999. DOI: 10.1145/336992.337035 Cited on page(s) 25

[174] T. Schiex, H. Fargier, and G. Verfaillie. Valued constraint satisfaction problems: Hard and easy problems. In *Proceedings of the Fourteenth International Joint Conference on Artificial Intelligence (IJCAI 95)*, pages 631–639. Morgan Kaufmann, 1995. Cited on page(s) 11, 14

[175] A. Schwenk. What is the correct way to seed a knockout tournament? *The American Mathematical Monthly*, 107:140–150, 2000. DOI: 10.2307/2589435 Cited on page(s) 52

[176] Y. Shoham and K. Leyton-Brown. *Multiagent Systems: Algorithmic, Game-Theoretic, and Logical Foundations*. Cambridge University Press, Cambridge, UK, 2009. Cited on page(s) 55

[177] X. Su and T. M. Khoshgoftaar. A survey of collaborative filtering techniques. *Advances in Artificial Intelligence*, 2009, 2009. DOI: 10.1155/2009/421425 Cited on page(s) 25

[178] A.D. Taylor. *Social choice and the mathematics of manipulation*. Outlooks Series. Cambridge University Press, 2005. Cited on page(s) 45

[179] C.-P. Teo, J. Sethuraman, and W.-P. Tan. Gale-Shapley stable marriage problem revisited: Strategic issues and applications. *Management Science*, 47(9):1252–1267, 2001. DOI: 10.1287/mnsc.47.9.1252.9784 Cited on page(s) 61, 65

[180] E. Triantaphyllou. *Multi-Criteria Decision Making Methods: A Comparative Study*. Kluwer, 2000. Cited on page(s) 27

[181] A. Tsoukiàs and M. Öztürk. Preferences on intervals: a general framework. In *Preferences*, volume 04271 of *Dagstuhl Seminar Proceedings*. IBFI, Schloss Dagstuhl, Germany, 2006. Cited on page(s) 32

[182] J. Uckelman and U. Endriss. Compactly representing utility functions using weighted goals and the max aggregator. *Artificial Intelligence*, 174(15):1222–1246, 2010. DOI: 10.1016/j.artint.2010.07.003 Cited on page(s) 26

[183] C. Unsworth and P. Prosser. A specialised binary constraint for the stable marriage problem. In *Proc. SARA 2005*, volume 3607 of *Lecture Notes in Computer Science*, pages 218–233. Springer, 2005. DOI: 10.1007/11527862_16 Cited on page(s) 68

[184] C. Unsworth and P. Prosser. Specialised constraints for stable matching problems. In *Proc. CP 2005*, volume 3709 of *Lecture Notes in Computer Science*. Springer, 2005. DOI: 10.1007/11564751_107 Cited on page(s) 68

[185] X.-H. Vu and B. O'Sullivan. Semiring-based constraint acquisition. In *Proceedings of the 19th IEEE International Conference on Tools with Artificial Intelligence (ICTAI 2007)*, pages 251–258. IEEE Computer Society, 2007. DOI: 10.1109/ICTAI.2007.160 Cited on page(s) 4, 25

[186] T. Walsh. Uncertainty in preference elicitation and aggregation. In *Proceedings of the 22nd National Conference on AI*, 2007. Cited on page(s) 56

[187] T. Walsh. Complexity of terminating preference elicitation. In *7th International Joint Conference on Autonomous Agents and Multiagent Systems*. IFAAMAS, 2008. DOI: 10.1145/1402298.1402357 Cited on page(s) 56

[188] T. Walsh. Where are the really hard manipulation problems? The phase transition in manipulating the veto rule. In *Proceedings of 21st IJCAI*, pages 324–329, 2009. Cited on page(s) 54

[189] T. Walsh. An empirical study of the manipulability of single transferable voting. In *Proc. of the 19th European Conference on Artificial Intelligence (ECAI-2010)*. IOS Press, 2010. Cited on page(s) 50, 54

[190] V.V. Williams. Fixing a tournament. In M. Fox and D. Poole, editors, *Proceedings of the Twenty-Fourth AAAI Conference on Artificial Intelligence (AAAI 2010)*. AAAI Press, 2010. Cited on page(s) 52

[191] N. Wilson. Efficient inference for expressive comparative preference languages. In *IJCAI*, pages 961–966, 2009. Cited on page(s) 18

[192] L. Xia and V. Conitzer. Determining possible and necessary winners under common voting rules given partial orders. In D. Fox and C.P. Gomes, editors, *Proceedings of the Twenty-Third AAAI Conference on Artificial Intelligence (AAAI 2008)*, pages 196–201. AAAI Press, 2008. Cited on page(s) 56

[193] L. Xia and V. Conitzer. Generalized scoring rules and the frequency of coalitional manipulability. In *EC '08: Proceedings of the 9th ACM conference on Electronic commerce*, pages 109–118. ACM, 2008. DOI: 10.1145/1386790.1386811 Cited on page(s) 53

[194] L. Xia and V. Conitzer. A sufficient condition for voting rules to be frequently manipulable. In *EC '08: Proceedings of the 9th ACM conference on Electronic commerce*, pages 99–108. ACM, 2008. DOI: 10.1145/1386790.1386810 Cited on page(s) 53

[195] L. Xia, J. Lang, and M. Ying. Sequential voting rules and multiple elections paradoxes. In D. Samet, editor, *Proceedings of the 11th Conference on Theoretical Aspects of Rationality and Knowledge (TARK-2007)*, pages 279–288, 2007. Cited on page(s) 59

[196] M. Zuckerman, A.D. Procaccia, and J.S. Rosenschein. Algorithms for the coalitional manipulation problem. In S.-H. Teng, editor, *Proceedings of the Nineteenth Annual ACM-SIAM Symposium on Discrete Algorithms (SODA 2008)*, pages 277–286, 2008. Cited on page(s) 50

[197] http://econ.core.hu/english/res/game_app.html, Collection of applications at the website of the Game Theory research group, Institute of Economics, Hungarian Academy of Sciences, 2011. Cited on page(s) 62

Authors' Biographies

FRANCESCA ROSSI

Francesca Rossi is a full professor of Computer Science at the University of Padova, Italy. She works on constraint programming, preference reasoning, and multi-agent preference aggregation. She has published over 150 papers on these topics, and she has edited 16 volumes between collections of articles and special issues. She has been conference chair of CP 1998, program chair of CP 2003, and conference organizer of ADT 2009. She will be program chair of IJCAI 2013. She is a co-editor of the Handbook of Constraint Programming, with Peter Van Beek and Toby Walsh, published by Elsevier in 2006. She has been the president of the Association for Constraint Programming from 2003 to 2007. She is a member of the advisory board of JAIR (where she has been associate editor in 2005-2007) and of the editorial board of Constraints and of the AI Journal, an associate editor of AMAI, and a column editor for the Journal of Logic and Computation. She is an ECCAI fellow.

K. BRENT VENABLE

K. Brent Venable is currently an assistant professor in the Dept. of Pure and Applied Mathematics at the University of Padova (Italy). Her main research interests are within artificial intelligence and regard, in particular, compact preference representation formalisms, computational social choice, temporal reasoning and, more in general, constraint-based optimization. Her list of publications includes more than 50 papers, including journals and proceedings of the main international conferences on the topics relevant to her interests. She is involved in a lively international scientific exchange and, among others, she collaborates with researchers from NASA Ames, SRI International, NICTA-UNSW (Australia), University of Amsterdam (The Nederlands), 4C (Ireland) and Ben-Gurion University (Israel).

TOBY WALSH

Toby Walsh was most recently acting Scientific Director of NICTA, Australia's centre of excellence for ICT research. He is adjunct Professor at the University of New South Wales, external Professor at Uppsala University and an honorary fellow of Edinburgh University. He has been Editor-in-Chief of the Journal of Artificial Intelligence Research, and of AI Communications. He is both an AAAI and an ECCAI fellow. He has been Secretary of the Association for Constraint Programming (ACP) and is Editor of CP News, the newsletter of the ACP. Like Francesca, he is one of the Editors of the Handbook for Constraint Programming. He is also an Editor of the Handbook for Satisfiability. He has been Program Chair of CP 2001, Conference Chair of IJCAR 2004, Program and Conference Chair of SAT 2005, Conference Chair of CP 2008, and Program Chair of IJCAI 2011.

Printed in the United States
by Baker & Taylor Publisher Services